Grid-Connected & Standalone Solar Systems Design

Phineas S. Malunjwa
(Msc – Electrical Eng.)

Shield Crest

ISBN 978-1-912505-31-9

MMXVIII

A CIP catalogue record for this book
is available from the British Library

Published by
ShieldCrest Publishing
Aylesbury, Buckinghamshire,
HP18 0TF
England
+44 (0) 333 8000 890
www.shieldcrest.co.uk

Contents

Table of Figures

Table of Tables

Acknowledgements

I would like to thank my colleague and friend Ludo Ndhlukula for encouraging me to write this book and for his great assistance in reproducing most of the drawings within it. I thank my family, especially my niece, Phathiwe Ntini, who has been an inspiration all the way to the completion of this book. I would like to thank Tricia Hopkins and Dr Eshetu Wondimagegne very much for encouraging me to write. Also, I would like to send my sincere gratitude to the staff at the Millennium Library and the Sprowston Library, both in Norwich City, for their wonderful assistance in helping me in using the computer.

I hope the users of this book will find its contents very useful.

Phineas S. Malunjwa (Msc - Electrical Eng.)

Introduction

The purpose of this book is to assist engineers and electricians in the design and installation of photovoltaic electric energy systems. At the beginning of this book (Chapter 1), I have briefly discussed PV cell technology and different types of PV cells. I have deliberately given emphasis to photovoltaic (PV) cell/module efficiency because land is generally limited, and so is the available space for PV modules. The generation of photovoltaic electrical energy, compared to conventional energy generation, requires more space. In Chapters 2 and 3, I have given some discussion of capturing (or harnessing) the sun's energy and estimation of the electrical energy output for a grid-tied PV system, respectively. Chapters 4 and 5 are dedicated exclusively to the grid-tied PV system. In Chapter 4, the main components and other components (balance of system components) of a grid-tied PV system are discussed, and the main things to look for in the selection of the components are given. Chapter 5 outlines the procedure in the design of grid-tied PV systems. Chapter 6 focuses on standalone PV systems. The main components of a standalone PV system, namely a PV battery and standalone PV inverter, are explained, and the battery charge controller and other balance of system components are also discussed. A brief description of various configurations of standalone and hybrid PV systems is given. Several worked-out examples of grid-tied PV system designs are given, and one worked-out example for a standalone PV system design is provided. In Chapter 7, a list of reference material is given, with some references also given elsewhere in the book.

1. Photovoltaic Cell/Module

1.1 PV Cell/Module Technology

The most important component of the photovoltaic (PV) module or solar panel is the solar (PV) cell. The most common commercially available PV cells are monocrystalline PV cells and polycrystalline PV cells. These cells are manufactured from pure silicon. The monocrystalline PV cells are created from wafers cut from cylindrical ingots of single crystalline pure silicon, and the polycrystalline PV cells are made from wafers cut from a block of pure silicon consisting of a large number of crystals. The silicon is found in quarts or sand. This is heated at very high temperatures to produce pure silicon, which is doped with boron (group 3 element), called the acceptors, (to form the **p-type** junction) and poured from a second container into moulds and cooled. During the cooling process, the pure silicon is formed into ingots of square blocks of polycrystalline silicon. To produce monocrystalline silicon, the quartz (raw silicon) is heated to very high temperatures to produce pure polysilicon or polycrystalline silicon, and the pure silicon is further subjected to high temperatures. The molten pure silicon is then doped with boron (to form the **p-type** junction) and gradually cooled and worked upon to produce cylindrical ingots of monocrystalline silicon. From ingots of square blocks of polycrystalline silicon or ingots of quasi-square single crystal silicon, wafers with a thickness of about 160μm to 190μm are cut and used for making PV cells. During the cutting process of the monocrystalline wafers, more silicon material is lost than during the cutting of polycrystalline wafers; this is because the single crystalline ingots (which are cylindrical) have to be chipped on opposite sides to make them quasi-square. The manufacture of the monocrystalline PV cell is more complex and consumes more energy compared to that of the polycrystalline PV cell. As a result, the monocrystalline PV module is more expensive than the polycrystalline PV module. During the formation of the crystalline silicon photovoltaic cell, the wafers are doped with impurity atoms, often phosphorous (group 5 element), also called donors (**n-type** material) and boron (**p-type** material), to form a positive and negative junction (p-n junction), creating an electrical field. A PV cell conducting grid is fixed and cell contacts are fixed at the front and back of the PV cell. A non-reflective material is thinly deposited on the front surface of the PV cell to reduce light reflection from the cell. Monocrystalline PV cells are usually textured to minimise light reflection. This improves the PV cell efficiency.

The PV cells are then connected in series or in series and parallel. For each string of PV cells, connections to bypass diodes are provided. Bypass diodes are connected in parallel, but with opposite polarity, to PV cells. The bypass diodes provide passage for current to bypass shaded, dirty or defective PV cells, as a result avoiding the creation of hot spots (local overheating). Hot spots can lead to destructive effects, such as melting of the solder, cracking of the glass or destruction of the PV cell. In principle a bypass diode should be connected parallel to each of the PV cells connected in series. However, in practice one bypass diode is usually connected in parallel to several PV cells in a string of cells. The bypass diodes are housed in the module junction box which is fixed at the back of the PV module.

The assembled and inter-connected crystalline silicon solar cells, are insulated and encapsulated in poly-plastic polymer, usually ethylene-vinyl-acetate (EVA), and they are covered with tampered glass on top (front) and below (back) and sealed. In order to reduce PV module weight, the back of the PV module may be covered with some Tedlar composite or Maylor (polyester plastic sheet) or ceramic plate and sealed. A weatherproof PV module junction box is fixed at the back of the PV module (see ***4.2.1 PV module junction box***).

Other solar cells available are thin-film PV cells. Examples of thin-film PV cells are amorphous silicon PV cells (a-Si), cadmium telluride (CdTe) PV cells, and copper indium gallium selenide (CIGS) PV cells. In the manufacture of thin-film PV cells the thin film source material used is vaporised and deposited to a thickness (depth) of about 1μm to 2μm on to a substrate, which may be heat-resistant glass, aluminium, or plastic material. That means the substrate may be flexible. Thin-film solar cells are flexible. The thin-film solar cells are doped with impurity atoms to form the p-n junction, and conducting grids and contacts are fixed. Additionally, a non-reflective material is sprayed on the front surface of the PV cell to minimise light reflection, and the back metal contact of the PV cell may be coated with reflective material to reflect unabsorbed light back to the absorber.

In the manufacturing process of thin-film PV cells, less energy and material is used compared to the energy and material used in the manufacturing process of crystalline silicon cells. In addition, the manufacturing of thin-film cells is much easier. Usually, both the PV cell and the PV module are produced on the same production line. The commercially produced PV modules from thin-film cells are lighter and cheaper. However, they are less efficient compared to crystalline silicon PV modules.

When light falls on the solar cell a voltage is created at the PV cell terminals. The size of this voltage depends on the type of material used in the manufacture of the solar cell. In the case of a crystalline silicon PV cell, this voltage is about 0.5 VDC to 0.6 VDC. The cells are then connected together in series (making a string of PV cells) to get the required voltage for the PV module, and in parallel to get the required current. The number of PV cells in a PV string determines the PV module open-circuit voltage (V_{OC}). The V_{OC} is independent of the light falling on the solar cells (see **Figure 2**). For a crystalline silicon PV module with 36 PV cells connected in series, the open-circuit voltage would be about 18 VDC to 21.6 VDC.

1.2 PV Cell/Module Efficiency

The maximum efficiency (percentage output at maximum power point) of the solar cell largely depends on the material used and how the cell has been treated during manufacture. For example, whether the PV cell is coated with non-reflective material or textured will have an effect on the efficiency of the PV cell; similarly, the size of the PV cell area occupied by the grid (grid formation) will have an effect. Only cells from the same material, similarly processed and treated during the manufacture, can be assembled together. For example, the efficiency of the monocrystalline silicon cell is slightly higher than the efficiency of the polycrystalline solar cell, by about 2% to 5% *. The efficiency of a thin-film amorphous silicon solar cell is even lower, by almost 10% to 12% than that of the monocrystalline solar cell.

Thin-film solar cells, of less material cost and higher voltages and efficiencies, are being developed and are already on the market, such as the triple-junction solar cell with an efficiency of 13.7%, from an alloy of amorphous silicon carbon and germanium, and the copper indium gallium selenide (CIGS) PV cell.

The efficiency of a photovoltaic module is less than the efficiency of the PV cell by about 2.5%. The efficiency of the PV cell is higher than the efficiency of the of PV module due to various factors such as blank PV cell areas around PV module edge, blank areas between PV cells, light reflection losses, and losses in PV cell connections.

*see Progress in Photovoltaics: Solar Cell Efficiency Tables (Version 48), first published 17th June 2016. Accessed February 2017.

1.3 PV Cell/Module Temperature

The increase in PV cell/module temperature decreases the PV cell/module voltage, and the decrease in PV module/cell temperature increases the PV cell/module voltage. In other words, the increase or decrease in PV cell/module voltage is inversely proportional to the change in PV cell/module temperature. However, the value of the PV cell (or module) current is almost independent of change of temperature. This is shown in **Figure 1**. Therefore, the temperature of a PV module/cell will have an effect on PV module/cell efficiency.

The efficiency of a PV cell/module will decrease with the increase in PV cell temperature. The temperature coefficient of a monocrystalline PV cell output power is about –0.45%, while that of a polycrystalline PV cell is about –0.5%, and that of a thin-film PV cell, for example amorphous silicon (a-Si) and copper indium gallium selenide (CIGS), is about –0.25%. Above 25^{0}C, the PV cell voltage and efficiency decrease, and below 25^{0}C the PV cell voltage and efficiency increase. In other words, with regard to the decrease in efficiency, for every 1^{0}C rise in temperature above the temperature at standard test conditions (see ***1.6 PV Cell/module Peak Power and Efficiency***) for a monocrystalline silicon PV cell/module the power output will decrease by about 0.45%; for a polycrystalline PV module the output will decrease by about 0.5%; and for a thin-film PV module the output will decrease by about 0.25%. This means that the amorphous silicon PV cells/modules will perform well at high irradiance levels while the monocrystalline and polycrystalline silicon cells/PV modules will not perform as well as the amorphous silicon cells/modules. This is because the thin-film PV cells have a larger band gap (the depleted zone between valance band and conduction zone) compared to the band gap of crystalline PV cells (see also ***1.5 PV Cell Fill Factor*)**. This means that, since the thin-film PV cells have a larger band gap compared to the band gap of crystalline PV cells, they have a lower temperature coefficient. Therefore, thin-film PV modules are preferred in hotter areas because they are less susceptible to temperature effects (see **Figure 1**).

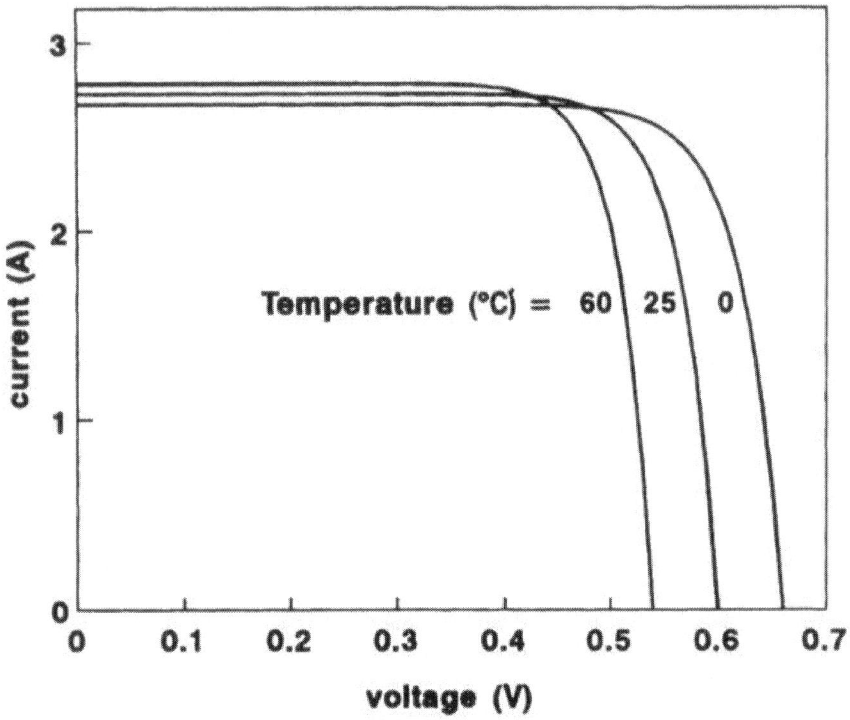

Figure 1: Effect of temperature on the current-voltage characteristics of a crystalline silicon solar cell *

*Source: http://www.itacanet.org/a-guide-to-photovoltaic-panels/photovoltaic-pv-cells/

(This was retrieved in February 2017)

The current is almost independent of change in temperature. That means, the circuit current (I) remains almost constant with the increase in PV cell/module temperature:

- There is substantial variation in open-circuit voltage (V_{OC}) – the PV (solar) cell/module voltage decrease is directly proportional to the increase in PV cell/module temperature, and vice versa. The increase in current with increase in cell temperature is negligible; likewise, the decrease in current with increase in PV cell temperature is negligible.
- The decrease in PV cell/module power at maximum power point (P_{MPP}) is directly proportional to the increase in PV cell/module temperature. The operating power of the cell/module occurs at the maximum power point, the point at which the PV cell/module power is equal to the product of the current at maximum power point (I_{MPP}) and the voltage at maximum power point (V_{MPP}). This is the maximum power, which the maximum power point tracker (of the grid-tied inverter searches in order to maximise the electrical power output of the PV array, and it corresponds to the knee of the I-V curve. Since the increase of temperature of a PV cell/module results in greater decrease in voltage, while the current remains almost constant, the PV cell/module power output is decreased.

Therefore, it is essential to make sure that PV modules are installed in such a way that there will be sufficient air flow below the PV modules. If PV modules are installed, where there is free air flow under the PV module, such as on flat roofs (on raised racks), or on the ground (such as pole mounted PV arrays), the PV module temperature will be about 20^0C to 25^0C above ambient temperature, and if modules are installed on pitched roofs or where there is some degree of restriction to the flow of air under the PV module, the module temperature can be as high as about 33^0C above ambient temperature*.

*See: **PDF The Energy Performance of In-roof PV – Viridian Solar, 2017.**

1.4 Effect of Irradiance on I-V of PV Cell/Module

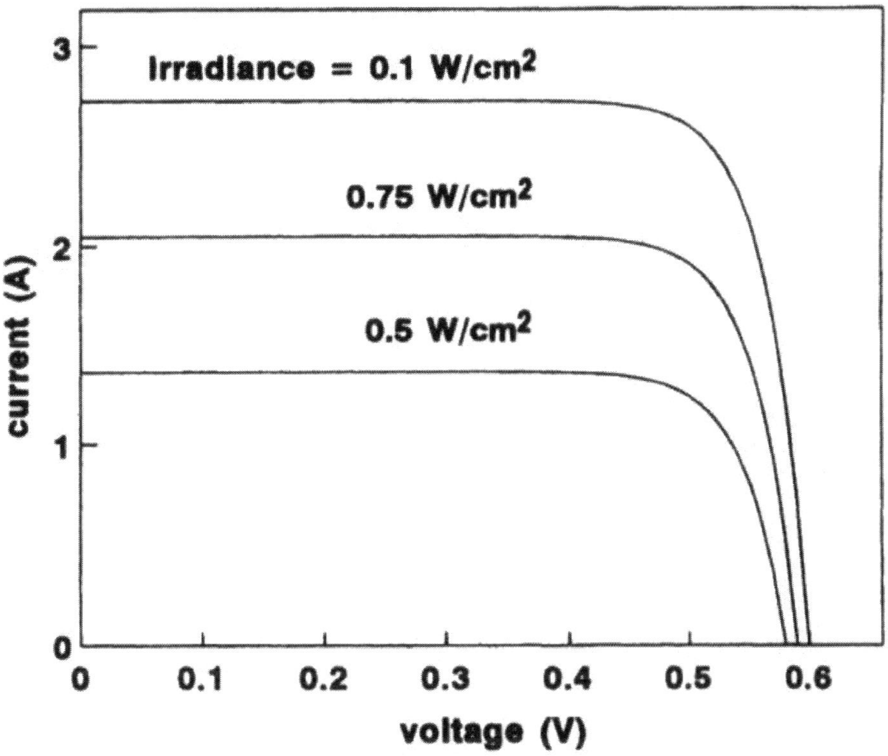

Figure 2: Current-voltage characteristic of PV module under varying levels of irradiance

*Source: http://www.itacanet.org/a-guide-to-photovoltaic-panels/photovoltaic-pv-cells/

(This was retrieved in February 2017)

Note the effect on electrical current, voltage, and power when solar irradiance on the PV module is increased.

Under varying irradiance:

- There is a proportionate increase in electrical current with the increase in irradiance (W / m^2) and a proportionate decrease in current with the decrease in irradiance. An increase in irradiance results in a proportionate increase of output current, i.e. if the irradiance doubles the current will also double.

- The open-circuit voltage is almost independent of irradiance. An increase in irradiance leads to a negligible increase in PV cell/module voltage (due to a slight increase in temperature).
- There is an increase in power output (W), and this increase in power is directly proportional to the increase in irradiance. Since the maximum power is at the knee of the current-voltage curve, the current at the PV cell's maximum power point (I_{MP}) is a little less than the short-circuit current, and the output voltage at the PV cell's maximum power (V_{PM}) is less than the open-circuit voltage (V_{OC}).

Under constant irradiance:

- The value of the PV cell/module current is independent of operating voltage.
- The value of the PV cell voltage will depend on the temperature, the type of material used in the manufacture of the PV cell, and on the number of PV cells/modules connected in series.

1.5 PV Cell Fill Factor (FF)

The ratio of the operating (actual) power ($V_{MPP}\ I_{MPP} = P_{MAX}$) to the product of the short-circuit current and the open-circuit voltage ($V_{OC}\ I_{SC}$) is referred to as the fill factor. So,

$FF = (V_{MPP}\ I_{MPP}) \div (V_{OC}\ I_{SC})$

$P_{MAX} = FF\ (V_{OC}\ I_{SC})$. The efficiency factor of a solar cell is the ratio of the PV cell output power (P_{MAX}) to the PV cell input power, determined at standard test conditions.

The fill factor of a PV cell is a measure of the efficiency of that PV cell, and it is always less than one. Most PV cells that are commercially produced have a fill factor greater than 0.7.

The output resistance of a PV cell at its maximum power point is the PV cell's characteristic resistance. If the resistance of the load is equal to the PV cell's characteristic resistance, the maximum power of the PV cell is transferred to the load and the cell operates at its maximum power point. The PV cell's characteristic resistance is the resistance the maximum power point tracker searches, enabling the PV cell to operate at its maximum power point. This resistance causes power losses.

PV cells with less internal series resistance and high shunt resistance have a higher fill factor because they have fewer internal losses (less of the current produced is dissipated by internal resistance). This means that the greater the PV cell series resistance (which is the resistance through the emitter and the base of the silicon, and between the silicon and metal contact and the top and lower metal contact of PV cell terminals), the smaller the PV cell fill factor. The size of the PV cell fill factor will also be determined by cell voltage (as a result of material used in the manufacture of the PV cell and PV cell temperature). The greater the band gap, which is the depleted zone between the valence band and the conduction zone, the greater the voltage of the solar cell.

1.6 PV Cell / Module Peak Power and Efficiency

The efficiency factor of a PV cell/module is the ratio of the PV cell/module maximum power output to the product of the irradiance and the PV cell/module surface area, and it is calculated at standard test conditions (STC).

$$\text{PV Cell Efficiency} = \frac{\text{Maximum power output}}{\text{Irradiance at STC} \times \text{ Area of PV cell (m2)}}\ \%$$

Since the irradiance at standard test conditions is equal to 1 kW / m^2 (see below), PV cell/module efficiency is therefore equal to the ratio of the PV cell/module maximum power output (in kW) to the PV cell/module area (m^2).

The electrical current produced and the electrical power output are directly proportional to the sunlight (irradiance) falling on the PV cell/module (see **Figure 2**). The peak power of the solar cell or module is obtained at maximum solar radiation; that is, it is achieved when there is maximum sunlight, at solar noon, during clear sky.

As has been explained, the electrical output of the PV cell or module varies with the amount of solar irradiance (W / m^2) falling on the PV cell/module and the amount of electrical power that the PV cell or module can produce is described as its peak power (W). The peak power of a PV cell or module is obtained at a standardised solar irradiance, and it is defined as energy output (watts) per square metre, obtained at 1000W per square metre (1000W / m^2) solar irradiance, at PV cell temperature of 25^{0}C and 1.5 Air Mass. The irradiance at standard test conditions is used for calibrating PV cells and modules. Air Mass is the optical path length, through the Earth's atmosphere, for light spectrum from a celestial source; alternatively, it can be described as a measure of the thickness of the atmosphere, which influences the spectral composition of the sunlight that reaches the Earth's surface due to the selective nature of the atmospheric gases. In other words, Air Mass is the relative length of the sun's direct beam path through the atmosphere. The distance between the sun and the PV module/array decreases as the sun moves toward zenith and increases as it moves toward the horizon and the sun's apparent movement across the sky similarly decreases and increases the distance. The increased distance leads to an increase in absorption gases, which change the spectral distribution of the irradiance (see ***2.1 Capturing of the Sun's Energy***). The actual operation of PV cell/module/array is generally at irradiance below the maximum solar irradiance of 1000W / m^2. For example, the actual operation of PV cell/array may be at a solar irradiance of as low as 850W / m^2. That means the PV array peak power is reduced by an irradiance factor of 0.85 (850W / m^2 divided by 1000W / m^2 = irradiance factor). Therefore, the irradiance factor should be taken into account when estimating a PV system's electrical energy output. In some locations, especially in deserts and tropical regions, irradiance may be much higher than the 1000W / m^2. The solar irradiance on a horizontal surface is made up of direct beam irradiance and diffuse irradiance (see ***2.1 Capturing of the Sun's Energy***).

When the sun is directly overhead, its zenith angle is equal to zero, its incident light is perpendicular to a point on the horizontal surface, and the Air Mass is equal to one at sea level (AM = 1). As the sun moves away from zenith, the Air Mass becomes greater than one. The longer the sun-path length, the more the scattering and absorption of solar radiation by atmospheric constituents, such as water and aerosols (air particles).

1.6.1 Example 1

Compare the energy output of two PV cells types, for example: first position, PV cell with surface area of 20cm x 20cm; second position, PV cell with a surface area of 10cm x 10cm; both PV cells with efficiency of 20% in the first position and 40% in the second position; and both PV cells operating under the same conditions, including latitude, climatic conditions, etc.

The PV cell/module maximum power output (W) can be calculated as follows: PV cell power output = area of PV cell irradiance at standard test conditions x PV cell efficiency.

Solution:

When the efficiency is equal to 20% (in the first position) for both PV cells, it will be calculated as follows:

(i) Power output of PV cell with surface area of 20cm x 20cm is equal to

$$(0.2 \times 0.2)\mathrm{m}^2 \times \frac{1000\ \mathrm{w}}{1\mathrm{m}2} \times \frac{20}{100} = 8\ \mathrm{W}$$

(ii) Power output of PV cell with surface area of 10cm x 10cm is equal to

$$(0.1 \times 0.1)\mathrm{m}^2 \times \frac{1000\mathrm{w}}{1\mathrm{m}^2} \times \frac{20}{100} = 2\ \mathrm{W}$$

When the efficiency is increased to 40% (in the second position) for both PV cells the power output will be calculated as follows:

(iii) Power output of PV cell with surface area of 20cm x 20cm is equal to

$$(0.2 \times 0.2)\mathrm{m}^2 \times \frac{1000\ \mathrm{w}}{1\mathrm{m}2} \times \frac{40}{100} = 16\ \mathrm{W}$$

(iv) Power output of PV cell with surface area of 10cm x 10cm is equal to

$$(\frac{10}{100} \times \frac{10}{100})\mathrm{m}^2 \times \frac{1000\mathrm{w}}{1\mathrm{m}^2} \times \frac{40}{100} = 4\ \mathrm{W}$$

Therefore, the peak power varies proportionally with the solar cell/module surface area; that is, the greater the surface area of the PV module/array, the greater its power output. This means that doubling the PV array surface area will double the PV array output. Modules with greater efficiency are smaller in surface area than modules with lower efficiency for the same peak power.

1.7 PV Module Physical Size and Weight

The area size of a PV module/cell is directly and largely influenced by:

- Efficiency: the higher the efficiency of the PV cell/module, the smaller the PV cell/module area for the same PV cell/module electrical energy output, and vice-versa (see ***1.6.1 Example 1***).
- The required voltage: the voltage of a PV cell varies with the type of material used in its manufacture, and the required open-circuit voltage, at standard test conditions, will require a certain number of PV cells to be connected in series.
- The need for easy handling during installation: the module must be easy to handle and lift during installation. Therefore, the number of PV cells that can be connected is limited by the physical size and weight of the module. The weight of the PV array installed on a roof must be such that it can be sustained by the roof. Therefore, the physical size of the PV array can be limited not only by available roof space and cost but also by the weight of the PV modules (see ***4.1.2 Selection of PV module for grid-tied PV system***).

1.8 Ensuring Continuity of PV Module /Array Power Output

A. To ensure continuity of the power output from the PV module/array the following must be observed:

- On the roof of a building, PV module frames should be fixed in a way that will give a minimum air gape of at least 25mm under the PV modules, to allow flow of air, in order to prevent over-heating of PV modules (see ***1.3 PV Cell/Module Temperature***).
- When laminates (PV modules without frames) are installed as part of the roof or where laminates are fixed on building façade, a flow of air below the PV modules should be allowed.
- On sites where temperature is expected to be very high, PV modules with a low temperature coefficient are preferred (see ***1.3 PV Cell/Module Temperature***).
- Near shading (partial shading) of PV module/s or PV array leads to reduced string power output. Any source of shading must be removed or avoided where possible, and where some shading is unavoidable, the fixing (laying out) of series and parallel PV modules must be carried out in such a way that there is the least shading of a PV string at any one time (see ***1.9 Partial Shading***). The PV modules/array must be installed clear of other installations or anything that is likely to cause any shading, and in cases where a PV module/array is installed on a flat roof, or on the ground, the PV modules must be sufficiently spaced to avoid the front row of PV modules shading the next row of PV modules just behind, or sufficiently raised to avoid being covered in snow or rain water and vegetation (when installed on the ground);
- PV modules/array must be regularly cleaned of dirt and dust, etc., which may reduce the

amount of sunlight reaching them.

B. Other factors that may affect the performance of the PV system:

- Higher module tolerances, e.g. a 100W PV module with tolerances of ±5% can produce 95W of electric power, and the PV array is likely to have higher mismatch losses. PV modules with lower tolerances, notably ±0% to ±3%, are preferred (see ***4.1.2 Selection of PV module for grid-tied PV system***).
- Mismatch of PV modules in PV string: this leads to power losses (PV modules in a string must be of the same type, have similar tolerances, and must operate under the same conditions, including irradiance, orientation, and tilt).
- Soiling or accumulation of dust and dirt on the PV modules/array: this can partially prevent sunlight from reaching the PV cell/module, leading to reduced PV cell/module/array electrical output and possible damage of PV cell/s or module/s.
- Inverter efficiency: the inverter must be compatible with the PV array and must be able to work efficiently on part-load and be suitable for the environment (check inverter technical data and manufacturer's instruction).
- Cable losses: these should be limited to less than 3% (check national codes and regulations).

1.9 Partial Shading

Any shading of PV module(s)/array should be removed or avoided as much as possible. Where a PV array is installed on building roof, it should be installed in such a way that PV modules are not partially shaded at any time by other installations, such as chimneys, television aerials, lightning spikes, electrical power lines, nearby buildings, or by nearby trees, etc. Where the partial shading cannot be avoided, the shading item should be repositioned or removed if possible. Where the partial shading item cannot be removed or repositioned, the partial shading of any PV string should be minimised as much as possible. The PV module layout should be such that the partial shading of any PV string, at any one time, is minimised as much as possible. Where the PV module/array is installed on a flat roof or on the ground, the PV modules or rows of modules must be sufficiently spaced to avoid shading (see ***1.8 Ensuring Continuity of PV Module/Array Power Output***).

The sun window (sky space), at least between 9 a.m. and 3 p.m. (for higher latitudes) or between 8 a.m. and 4 p.m. (for lower latitudes), should be free of any shading object. This means that the sun path (sky window) should be examined to find out the sources of shading, which can include building structures or high ground, such as hills or mountains. This can be done by using a path finder or other similar device. The information obtained can be used to determine the shade factor, which should be used in determining the PV array's electrical energy output. There are websites that give sun-path diagrams for various latitudes.

When a PV array is laid on a flat roof or on the ground the distance between modules must be such that the front row of PV modules does not cast a shadow on the immediate back row of PV modules. The distance between rows of flat roof-mounted PV array may be obtained from the installer of the roof or calculated. For a flat roof or ground-mounted PV array, the horizontal distance between the front row and the back row of the array should be calculated (for the winter solstice, when shadow is longest, i.e. when the sun is at its lowest elevation angle and narrowest azimuth angles) between 9 a.m. and 3 p.m. or between 8 a.m. and 4 p.m. The higher the latitude the longer the shadow; the taller the rows the longer the shadow; and the greater the sun window (sky space), for example 8 a.m. to 4 p.m. for the same latitude, the longer the shadow; the opposite would be vice versa, respectively.

1.10 PV Module Set-Up and PV Array Power Output

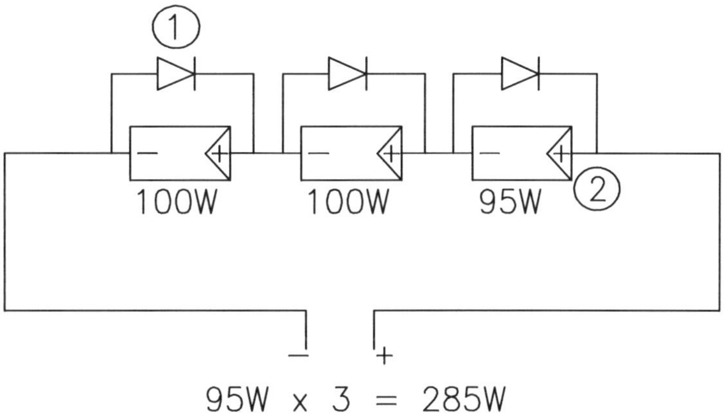

REF	DESCRIPTION
1	BYPASS DIODE
2	PV MODULE

Figure 3: PV modules connected in series

For modules connected in series, the PV string current remains the same as the current for one PV module in the PV string. In fact, the PV string current is equal to the lowest PV module current in the string. The string voltage is the sum of the voltages of all the PV modules forming the string. So the total power of each string is the lowest PV module power in the string multiplied by the number of PV modules in the string.

In a PV string, individual PV modules should be of the same type, and should have the same power rating and the same power tolerance as each of the other PV modules in the string. Additionally, the irradiance on one PV module should be the same as the irradiance on each of the other modules in the PV string. Therefore, the PV modules in the string should be of the same type, i.e. they should have the same electrical characteristics, and they should have the same orientation and tilt angles. Where this is not the case there will be mismatching, resulting in power losses. In addition, where there is unavoidable partial shading the PV modules in a string should be positioned to minimise this shading (see ***1.9 Partial Shading***).

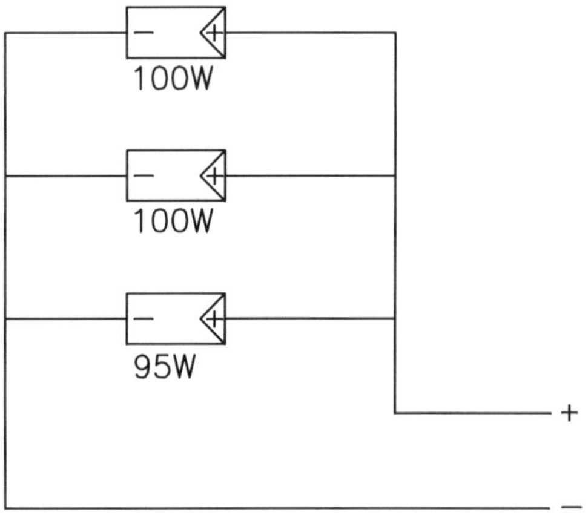

95W + 100W + 100W = 295W

Figure 4: PV modules connected in parallel

When PV modules are connected in parallel, the PV array current is the sum of the currents of all the PV modules or PV strings connected in parallel, and the voltage of the PV array remains unchanged (PV module voltage or PV module voltage multiplied by the number of PV modules connected in series). This means that the PV array power is equal to the sum of powers (W) of the parallel PV modules or strings. In Figure 5, below, the total wattage is determined as follows:

100W × 3 = 300W (wattage of first string)

95W × 3 = 285W (wattage of second string)

Total PV array power = 300W + 285W = 585W

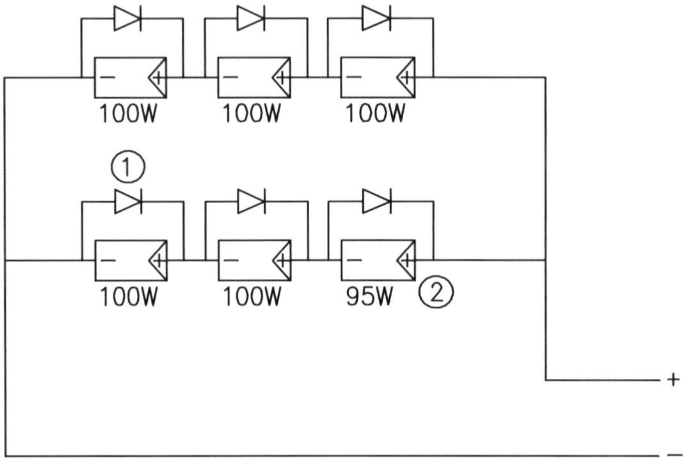

Figure 5: PV modules connected in parallel and series

Bypass diodes are provided, if required (consult PV module manufacturer), to enable the current to bypass shaded, dirty or defective PV cells (also see **Figure 6, Figure 10** and **Figure 11**).

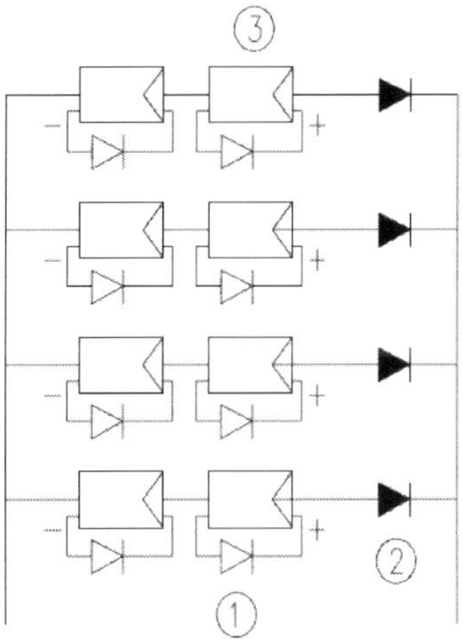

REF	DESCRIPTION
1	BYPASS DIODE
2	BLOCKING DIODE
3	PV MODULE

Figure 6: PV modules connected in parallel and series

In **Fig. 6**, the blocking diode will prevent the flow of current from a healthy PV string into a shaded, or failed PV string. The blocking diode can be replaced by a string fuse (consult module manufacturer, national codes and regulations). In standalone PV systems the blocking diode will prevent current from the battery into the array (see **Figure 10** and **Figure 11**). Blocking diodes are less efficient compared to fuses, they require regular testing, and they are less reliable because they are subject to frequent breakdown.

2. The Sun's Energy

2.1 Capturing of the Sun's Energy

Solar irradiance, which is the solar radiation in kW / m^2 received by a specific surface or projected on to a specific surface, is most abundant at solar noon during a clear sky. In order to capture the most sunlight (solar irradiance) and the maximum amount of direct beam irradiance, the PV module/array should be placed perpendicular to the sun's rays, being oriented toward the south for locations in the northern hemisphere and toward the north for locations in the southern hemisphere.

The sun's altitude (its elevation angle to the horizon) changes with the latitude, day in the year, and solar time, as the Earth rotates on its tilted axis and moves along its ecliptic path around the sun. The Earth's tilt on its polar axis, relative to the plane of its movement around the sun, results in the sun's declination angle. This angle, which is the angle measured between the centres of the sun and the Earth and the plane of the equator, varies with the day in the year and the solar time, and it varies between $+23.45^0$ (at the summer solstice) and -23.45^0 (at the winter solstice). This means that the sun's declination at any location on the Earth's surface cannot be greater than 47^0. At lower latitudes, at solar noon, the sun will be almost overhead, and it will be directly overhead at some point during the year at solar noon, between the tropics of Cancer and Capricorn at equinoxes (on 23 March and 22 September). At equinoxes, the sun's declination angle is equal to zero; hence, it will rise and set east and west, respectively, everywhere on the Earth.

The sun's elevation angle, at solar noon, is lowest during the winter solstice and highest during the summer solstice. In the northern hemisphere, the sun is at its furthest point to the south of the equator at the winter solstice (Tropic of Capricorn) on 21 December. The sun rises in the south-east and sets in the south-west during the winter months. As a result, in winter, the sun travels a shorter distance and further (to the horizon) across the sky. This means that a PV module/array will collect less of the sun's energy in winter. In summer, in the northern hemisphere, the sun is at its furthest point to the north of the equator at the summer solstice (Tropic of Cancer) on 21 June. The sun rises in the north-east and sets in the north-west. Therefore, the sun travels a longer distance across the sky during the summer months. Consequently, a PV module/array will collect more of the sun's energy each day in summer than it does in winter. In the southern hemisphere, the sun is at its furthest point to the north of the equator at the winter solstice (Tropic of Cancer) on 21 June. The sun rises in the north-east, travels away from overhead, and sets in the north-west. Therefore, the sun travels a shorter distance across the sky during the winter months. In summer, the sun is at its furthest point to the south of the equator at the summer solstice (Tropic of Capricorn) on 21 December. The sun rises in the south-east, travels up the sky to almost overhead, and sets in the south-west. Therefore, it travels a longer distance across the sky in the summer months than it does in winter. This means that a PV module/array will collect more of the sun's energy in summer than in winter. In higher latitudes, the sun travels a still shorter distance across the sky in winter than in summer, and, as a result, the difference between hours of sunlight per day for summer and for winter becomes much greater. As mentioned earlier, the sun's elevation angle varies with the latitude, the day in the year, and the solar time.

A PV module/array installed at a fixed-tilt angle should be tilted toward the sun's average elevation angle in order to collect the maximum amount of direct beam solar radiation.

The sun's average elevation angle at solar noon (for the whole year) is equal to the local latitude angle (see ***2.3 PV Module/Array Optimum Tilt and Orientation Angles***). This means that PV modules should be tilted at a fixed angle of latitude for the whole year in order to produce maximum electrical energy output. However, the optimum tilt angle may be different from latitude tilt due to other factors (see ***2.2 Other Factors that Determine PV Array Electrical Energy Output***).

The azimuth angle, which is the angle measured from north to any point on the horizontal perpendicular to the sun, will vary with latitude, day in the year, and solar time. The variation of the sun's azimuth angle is due to the Earth's tilt ($\pm 23.45^0$) on its axis of rotation, the sun's apparent movement along the ecliptic path around the Earth, and the local solar time. At equinoxes, the sun's azimuth angle at sunrise is 90^0 and at sunset it is 270^0. This means that the sun rises directly in the east and sets directly in the west.

As already explained, in the northern hemisphere, the sun is in the south while in the southern hemisphere it is in the north. Therefore, in locations in the northern hemisphere regions, the orientation (the azimuthal angle between the north and the azimuthal component of the normal to the plane of the PV module/array) of a PV module/array should be true south in order to produce maximum electrical energy output, while in locations in the southern hemisphere regions a PV module/array should face true north. If true south (in the northern hemisphere) or true north (in the southern hemisphere) is obstructed by ground rise or mountain or building, the PV module/array may be turned to face south-east or south-west of the local meridian (in the northern hemisphere) and east of north or west of north (in the southern hemisphere). Studies have shown that a small deviation from true south or true north has very little effect on PV module power output. In fact, up to 30^0 of deviation from true north or south will result in less than 5% loss of optimum PV module energy output. For example, in **Figure 7**, with a PV array oriented 30^0 east of south and located at latitude 40^0 the power output is 95% of optimum power output for a PV array that is optimally tilted and oriented due south.

The main factors that contribute to the amount of electrical energy output of a PV module/array, as previously described, are the latitude, the sun's declination, the sun's azimuth angle, the PV module/array tilt angle, and the PV module/array orientation. However, there are other factors that will have an effect on the electrical energy output of a PV module/array (see **2.2** ***Other Factors that Determine PV Array Electrical Energy Output***).

The solar irradiance (kW / m^2) striking on a horizontal surface, in locations with humid climates, such as locations in higher latitudes and in equatorial regions, has less direct irradiance and more diffuse irradiance because most of the direct irradiance is scattered by dust and air particles in the atmosphere, humid air, and foggy or cloudy weather conditions. This also means that installing a tracking system for your PV array would not be worthwhile. A tracking system is a system that keeps the PV array perpendicular to the sun by following the sun's apparent movement. As stated earlier, in order to capture most of the direct irradiance (kW / m^2) the PV module/array must be tilted perpendicular to the sun's rays. It follows, therefore, that in locations where there is more diffuse irradiance than direct irradiance, the optimum tilt angle of a PV module should be less than latitude. This is so in order to capture more of the diffuse irradiance. The diffuse irradiance is at its maximum when it is striking on a horizontal surface. Tilting the surface from the horizontal reduces the amount of diffuse irradiance falling on the surface, because a tilted surface from the horizontal will only see part of the sky dome.

Solar irradiation (kWh /m^2) – which is total irradiance integrated over a period of time – received by a horizontal surface or on a tilted surface is referred to as global irradiation (global radiation). Global irradiation on a tilted surface is made up of direct beam irradiation, scattered irradiation (diffuse), and reflected irradiation (B Y H Liu and R C Jordan, 1960). For example, Jimenez and Castro (1982) suggested 80% direct beam irradiation and 20% diffuse irradiation. The scattered sunlight is due to air particles (aerosols) present in the atmosphere, and the reflected sunlight depends on the ability of the ground surrounding the PV module/array to reflect sunlight, a property expressed by the albedo factor. The albedo factor for various surfaces can be found in the literature; for example, the albedo for naked ground in a humid, hot tropical location is 0.2, for naked ground in a dry tropical location it is 0.5, and for ground covered with snow it is 0.9 (T. Muneer 2004). On a tilted PV module/array, the reflected irradiance is a component of both the direct irradiance and diffused irradiance multiplied by the albedo factor. There are various ways of determining the diffuse component of the irradiance depending on the sky cloud cover (on the sky dome). One is isotropic, which assumes that there are no obstacles on the horizon, such as mountains or buildings, and that clouds of uniform brightness cover the sky (models by: B. Y. H. Lui and R. C. Jordan, 1960; P. S. Koronakis, 1986; V. Badescu, 2002; and others). The other is anisotropic, which assumes that there are obstacles on the horizon and that clouds of varying brightness cover the sky (models by: J. E. Hay and J. A. Davies, 1980; D T Reindl et al., 1990).

As can be seen from the above discussion, calculation of the diffuse irradiation on a tilted surface is more complex and can be time consuming. However, it is easier to calculate the direct beam irradiation on the tilted surface and then subtract the direct beam on the tilted surface from the global irradiation on the horizontal surface in order to obtain the diffuse irradiation.

Diffuse radiation = global radiation (on horizontal surface) – direct beam radiation on tilted surface.

2.2 Other Factors that Determine PV Array Electrical Energy Output

Other factors that affect PV array energy output include (see also ***1.8 Ensuring Continuity of the PV Module/Array Power Output***):

i) Weather conditions: local weather conditions, with cloudy or foggy days, result in less direct sunlight and more diffuse sunlight. In such locations, such as in most northerly locations in the northern hemisphere, the optimum PV array tilt angle can be found to be much lower than the location's latitude in order to capture the most of the diffused sunlight. Where there is frequent snow, a balance must be reached between capturing as much of the diffused radiation as possible and raising the PV array tilt angle in order to prevent partial shading of the PV array by the snow. A raised tilt angle will cause the snow to slide off the PV module.

ii) Soiling and dust pollution: accumulation of dust on a PV array reduces energy output by about 5% (Photovoltaic Modules: Effect of Tilt Angle on Soiling by Jose Cano (2015)). Rainfall of about 5mm per day, once a month, is enough to satisfactorily clean a tilted PV module/array. The PV array at higher latitudes (greater PV array tilt angle) can be better cleaned by the rain than those at lower latitudes when a PV array is tilted at latitude angle. At lower latitudes, or at locations with little rainfall, say less than 5mm once a month, it may be necessary to increase the PV array tilt angle above latitude in order to reduce soiling and dust pollution and to assist in the washing down of grime and dust from the PV array. In dry locations, or locations with less than 5mm of rainfall once a month, or to remove birds' droppings, it is necessary to implement an artificial wash-down and proper cleaning (in the case of birds' droppings). Regular cleaning of the PV module/array, apart from helping in maintaining PV module/array power output, prolongs the PV module's lifespan and prevents damage of the module(s). For a PV module/array with a tilt angle equal to between 0^0 (horizontal) and 23^0, regular cleaning of the modules should be carried out.

iii) Partial shading of PV module/s will reduce PV module/array output. Any partial shading of PV modules should be avoided or the shading objects should be removed where possible (see ***1.9 Partial Shading***). Dry leaves, from vegetation, that fall and remain on PV modules do shade some of the PV cells. The dry leaves and any dirt on the PV module should be removed regularly.

iv) Industrial pollution: PV modules/arrays should not be installed in areas where waste emissions can cloud the solar window and or pollute the PV module(s).

v) Temperature: increased PV module temperature will reduce module power output (see ***1.3 PV Cell/Module Temperature***).

2.3 PV Module/Array Optimum Tilt and Orientation Angles

Due to the sun's declination, the (average) optimum PV module/array tilt angle (tilt angle from the horizontal) should be equal to the site latitude minus the sun's declination angle. This is true if we consider the direct beam irradiance (kW / m^2) only. Under this consideration, at a specific latitude, the average fixed optimum tilt angle for summer will be equal to the latitude minus the sun's declination, and the fixed optimum tilt angle for winter will be latitude plus declination angle. The fixed optimum PV module tilt year-round should be equal to the site latitude. However, if we take into consideration dust and air particles present in the atmosphere, local weather conditions, and variation of the sun's distance from the earth, the size of tilt angle may change.

After considering the effects of air particles present in the atmosphere, climatic conditions, and the variation of the sun's distance from the Earth, it is standard that the PV module/array may be fixed at an optimum tilt angle of latitude -15^0 for summer or $+15^0$ for winter. J. A. Duffie and W. A. Beckman (1991) suggest a fixed optimum tilt of latitude $+10^0$ to $+15^0$ during winter and of latitude -10^0 to -15^0 during summer. B. Y. H. Lui and R. C. Jordan (1962) suggest a fixed optimum tilt of latitude $\pm10^0$ during winter and summer, respectively. **V. Heywood (1971)**, H. P. Garg (1982), and G. Lewis (1987) suggest a fixed optimum tilt of latitude $+8^0$ to $+15^0$ during winter and -8^0 to -15^0 during summer. The above values of optimum tilt angles may require adjustment depending on the local weather conditions. If the site location is often cloudy or foggy, the tilt angle may be lowered so that the PV module/array can collect most of the diffuse irradiation.

The fixed optimum PV module/array tilt angle for the whole year is equal to latitude, as explained earlier. The fixed latitude tilt angle, which results in the greatest yearly energy output, is used in most grid-tied PV systems, more so in tropical and desert areas, where there is abundance of direct sunlight. The summer average optimum PV array tilt angle for the whole year may be suitable where most energy is required in summer; for example, where electrical energy is needed for pumping water for irrigation, livestock watering, and cooling or refrigeration. The winter average optimum tilt angle year-round is suitable where winter energy requirements are essential. In other words, the summer average optimum tilt year-round is adopted where the main loads are in summer, and the winter optimum tilt year-round is adopted where the main loads are in winter. Both the summer optimum tilt year-round and the winter optimum tilt year-round are suitable for standalone PV systems (see ***6.3 Solar Resource for Standalone PV Systems***). The summer optimum tilt year-round should be used in grid-tied PV systems to maximise annual output of the PV modules/array. The fixed PV array tilt angle may be the average optimum PV array tilt angle adjusted twice a year, in which there is fixed average optimum tilt for summer and fixed average optimum tilt for winter. In addition, the fixed average optimum tilt angle may be adjusted each season, i.e. average optimum tilt for summer, average optimum tilt for autumn, average optimum tilt for winter, and average optimum tilt for spring. This arrangement can improve the output of a PV system and can be used in both the grid-tied and standalone PV systems. However, adjusting PV array tilt comes with associated extra costs, as we shall see below.

Since the sun's irradiance (kW / m^2) varies with its apparent movement from sunrise to sunset, the location's latitude, the day in year, and solar time, a PV system that tracks the sun may be required in order to increase electrical energy output. However, in locations in the higher latitudes and in humid regions it may not be advisable to install a PV array tracking system. In the higher latitudes the weather is mostly cloudy and humid, the sun travels low in the horizon in winter, and the hours of daylight are much less in winter. Therefore, the tracking system will collect less of the direct irradiation. Additionally, as already stated, in cloudy and humid areas most of the solar irradiation is mostly diffuse irradiation. The purpose of installing a tracking PV system is to collect as much as possible of the direct beam component of the sunlight.

There are formulas for calculating PV module optimum tilt angle for a specific day in the year. The tilt angles determined for specific days in the year are mainly very useful for standalone PV systems. These formulae can easily be found in the literature. There are also some websites that give factors for calculating fixed optimum PV module tilt angle for a whole year, or fixed optimum tilt adjusted twice or four times a year (seasonally), for specific latitudes. These also give dates on which the adjustments of optimum PV module tilt angle for each season should be made. However, it is important to remember that the optimum tilt angles obtained above may require adjustment to take into account the effects of local weather conditions. The fixed optimum tilt angle adjusted seasonally is suitable for grid-tied PV systems, and usually for small standalone PV systems. For example, **Table 1** below compares percentage of optimum output for a PV system, at various tilt, installed at latitude 40^0. The table is taken from the work of Charles R. Landau, 2015, (note that the comparison would be different for different latitudes).

Table 1: Percentages of optimum output for a PV system, at various tilt, installed at latitude 40^0

	Fixed year-round	Adjusted two seasons	Adjusted four seasons	Two-axis tracker
% of optimum	7.1%	75.2%	75.7%	100%

Adjusting the optimum tilt angle twice or four times a year, or axis tracking (where the PV array tracks the sun), gives greater electrical energy output from the PV module/array. However, this comes with associated extra costs. For example, a PV array installed at fixed optimum tilt angle year-round will produce 71.1% of optimum energy. Adjusting twice yearly gives an increase in energy output of 4.1% (75.2% instead of 71.1%), adjusting four times a year gives a further increase of only 0.5% (75.2% instead of 74.7%), while a two-axis tracker gives 100% of optimum energy. The choice of which system to adopt will depend mainly on cost, size, and energy requirement. Adjusting twice yearly has the added cost of the mounting structure, but has a 4.1% increase in energy yield compared to the fixed optimum tilt year-round; adjusting four times yearly has a further increased cost on mounting structures and a slight further increase (0.5%) in energy yield; the two-axis tracker can give 100% of optimum energy yield. In all the above cases of adjusting the PV module/array optimum tilt angle, there is additional cost involved. Adjusting seasonally and two-axis tracking leads to mounting structures with higher cost, and the moving structures are subject to wear and possible breakdown. In addition, electrical energy may be required, for moving the supporting structures or in adjusting the tilt angle (where the adjusting is not done manually). If the PV system is small, a one-axis tracker or even two-tracker system can be considered, since the energy yield is likely to outweigh the cost of extra supporting structures and maintenance. Similarly, if energy requirement is the main factor and extra supporting structure costs and maintenance costs are manageable, adjusting twice yearly could be the best option, since there will be a substantial increase of energy yield of 4.1%. Adjusting four times yearly is not advisable to adopt, unless there are strong reasons to do so for the extra 0.5% of optimum energy compared to a percentage of 75.2% of optimum energy for adjusting twice yearly. Therefore, choosing the adjusting of optimum tilt angle seasonally or axis tracking should be properly weighted against overhead costs in order to avoid greatly increasing the payback period of the PV system.

As has been indicated, there are two types of PV array axis tracking systems: single-axis and two-axis tracking systems. A single-axis tracking system varies the azimuth angle only, in order to track the movement of the sun across the sky, while a two-axis tracking system varies both the sun's azimuth and the sun's altitude angles in order to track the movement of the sun across the sky, from east to west. Comparing the electrical energy output from the two systems, it is suggested that a single-axis tracking PV system, with fixed optimum tilt angle, will produce an energy output of around 30% more than a non-tracking system (with fixed optimum tilt angle), while a two-axis tracking system will produce only about 5% more in energy output than the one-axis tracker. This weighs against the adoption of a two-axis tracker because of possible higher extra overhead costs for a mere 5% extra energy output. Axis tracking PV systems are applicable mainly where maximum energy is required from the PV system and preferably in the tropical regions, or in latitudes not greater than 30^0. In higher latitudes, because the sun travels a shorter distance across the sky and the weather is often cloudy and humid, the axis tracking system is not suitable. In larger PV systems, axis tracking can lead to unsustainable overhead costs, as indicated earlier.

The PV array, for most grid-tied PV systems, is installed flat on an existing building's pitched roof, or as part of the building's roof structure or on the building's façade. That means that the PV array's tilt angle and the orientation angle have already been determined by the building's roof pitch and orientation; the same applies for PV arrays installed on façades. However, it is suggested that deviation from the optimum tilt angle by 10^0 to 15^0 may lead to less than 5% loss of optimum energy output (see **Figure 7**). Locations with humid climates, like most locations in most northerly latitudes, have an optimum tilt angle of less than latitude, e.g. Hamburg, Germany has an optimum PV module tilt angle of 30^0 (see ***2.4 Example 2***). This is because in these locations the diffuse irradiation is much higher; in fact, in most northerly regions the diffuse radiation is about 50% of the total irradiation because the weather is often cloudy and humid, as explained earlier.

Locations in lower latitudes with a less humid climate, and locations in desert areas, have a direct component of solar radiation that is much higher than the diffuse component, as previously explained. PV modules installed in these locations should be tilted toward the sun's rays so as to capture most of the direct component of the solar radiation. In other words, PV arrays should be tilted perpendicular to the sun's rays. In lower latitudes, some optimum tilt angles tend to be too shallow; therefore, it may be necessary to raise the tilt angle in order to assist in cleaning dust and dirt from the PV array (see ***2.2 Other Factors that Determine PV Array Electrical Energy Output***). The raising of the PV module/array tilt angle will make the cleaning of the modules easier. It has been suggested that a minimum tilt angle of 12^0 is adequate for PV array cleaning.

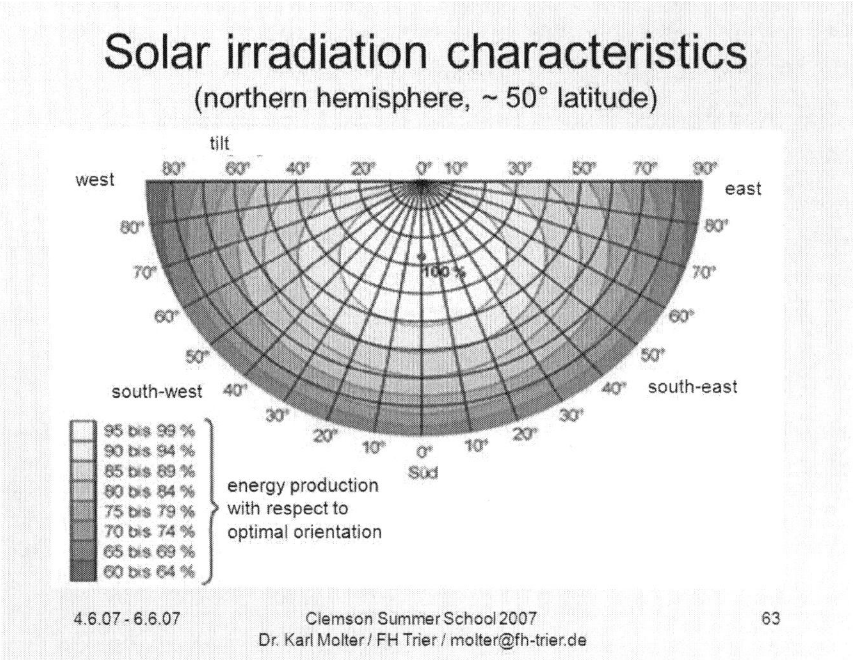

Figure 7: Percentage power output.

Figure 7 is a contour (yield) diagram of the global irradiation averaged over the year for specific locations, plotted for different PV module tilts and orientations. The colour bar represents sun-hours.

2.4 Example 2 – Radiation on a Tilted Surface

In this example, **Tables 2**, **3**, and **4** show the calculated percentage yield for various PV array tilt angles for locations in Sri Lanka, Lesotho, and Hamburg, respectively.*

i) Sri Lanka, which is at latitude 6^0 north of the equator, gives the following table:

Table 2: Sri Lanka: array tilt angle and percentage of optimum output

Array Tilt Angle	% of Optimum Output
0^0	99%
6^0 (Optimum Tilt Angle)	100%
15^0	100%
20^0	99%
25^0	97%
30^0	95%
35^0	92%
45^0	85%

ii) Lesotho, which is located at latitude 30^0 south of the equator, gives the following:

Table 3: Lesotho: array tilt angle and percentage of optimum output

Array Tilt Angle	% of Optimum Output
0^0	89%
10^0	96%
30^0 (Optimum Tilt Angle)	100%
35^0	99%
45^0	96%
50^0	94%
55^0	91%
60^0	87%

iii) Hamburg, located at latitude 53^0 north of the equator, gives the following:

Table 4: Hamburg: array tilt angle and percentage of optimum output

Array Tilt Angle	% of Optimum Output
0^0	87%
20^0	98%
25^0	99%
30^0 (Optimum Tilt Angle)	100%
45^0	99%
50^0	98%
55^0	96%
60^0	94%

In Sri Lanka, changing from the 6^0 optimum tilt to a 20^0 tilt angle results in a **1%** loss of optimum energy output, and changing the optimum tilt angle of 6^0 to 35^0 results in **8%** loss of optimum energy output.

In Lesotho, changing from the 30^0 optimum tilt to a 45^0 tilt angle results in a **4%** loss of optimum energy output, and changing the optimum tilt angle of 30^0 to a 60^0 tilt angle results in a **13%** loss of optimum energy output.

In Hamburg, changing from the 30^0 optimum tilt to a 45^0 tilt angle results in a **1%** loss of optimum energy output, and changing the optimum tilt angle of 30^0 to a 60^0 tilt angle results in a **6%** loss of optimum energy output. Note that the optimum tilt angle for PV array in Hamburg is much less than the latitude. This is because Hamburg, Germany, has a cloudy and humid climate and, as a result, there is an increased amount of diffuse solar radiation (about 50% of total radiation). There are similarities in energy losses for the same increase of tilt angles for the locations in Sri Lanka and in Hamburg because the two places have similar weather conditions (cloudy and humid). In Lesotho, most of the solar irradiation is direct irradiation. Note that with a change of up to 15^0 from the optimum PV array tilt angle the power loss from the optimum power output is less than 5%.

The locations with humid climates have optimum tilt angle less than latitude by about 10% to 25%.

*Note: The tables were made by using calculations from Green Rhio Energy; Solar Radiation on Tilted Surface, (http://greenrhioenergy.com/solar/radiation/tiltedsurface.php –retrieved on 09/09/2017).

2.5 Solar Resource

Solar radiation (kWh / m^2 / day) is usually given as global irradiation on a horizontal surface for specific regions or locations, or as global irradiation on an optimally tilted surface for specific regions or locations.

The sun's irradiance, on a specific surface and averaged or integrated over a given time, is global irradiation (kWh / m^2) (see ***2.1 Capturing of the Sun's Energy***). Global irradiation is equivalent to the average peak sun-hours (see ***1.6 PV Cell/Module Peak Power and Efficiency***).

The accuracy in the determination of solar irradiation (kWh / m^2 / day) that will be used will greatly determine the output of the PV system. Incorrect (under- or over-valued) solar radiation data can result in the PV system failing to work to expectation. The best solar radiation data is that collected from local sources, because it reflects local environment and weather conditions. Check with the local university, government institutions (agencies) that deal with weather conditions, and local newspaper organisations to see if they have available radiation data that can be used for the PV system design. The available information could be in the form of irradiation or irradiance data, collected (using a pyranometer or a pyrheliometer, respectively) for over a year or more. If the measured irradiance (solar radiance flux density, kW / m^2) is for over a year or more it can be calculated and averaged per day. If the global solar radiation data has been collected at specific tilt and orientation, check orientation and tilt angles of your PV system to make sure that the data will be suitable. There are also websites that give solar irradiation data, often in the form of maps, bar charts, tables, or graphs. The irradiation given can be for average year or for each month or for each day in a given month, on a horizontal or optimally tilted and oriented surface. The average peak sun-hours per day, at specific tilt and orientation angles, are most suitable for standalone PV systems. The hourly irradiation data is suitable for grid-tied PV systems where more electrical energy is required to be fed into the grid at a specific time of the day. There are also charts that show the percentage of maximum solar yield (average daily peak sun-hours) at various latitude tilts and orientation for specific locations (see **Fig.** 7). Although these charts may not be useful for the estimation of energy output for larger PV systems some may be useful for estimating smaller PV systems' energy output for grid-connected PV systems, because the error that can arise from the use of these charts in determining the PV energy output is likely to be negligible. Solar resource can also be given as kWh / KW at specific tilt and orientation of PV array, and this is suitable for estimating the energy output for grid-connected PV systems (see ***5.3 Example 6***). It should also be noted that the sun-hours for your latitude may need to be adjusted to take into account local conditions, e.g. foggy weather, industrial pollution, etc. In making the adjustment of sun-hours, to allow for local conditions, it is suggested that due care must be taken not to exceed 20% of global solar irradiation. Therefore, average sun-hours data obtained from solar radiation maps, charts, graphs, or tables for particular locations are most valuable. Some of the websites from which irradiation data is obtainable are: http://weathersa.co.za (South Africa), http://eosweb.larc.nasa.gov/ NASA Surface Meteorological and Solar Energy Data Set), http://re.jrc.ec.europa.eu/pvgis/, (PVGIS EU Joint Research Centre Solar radiation data and maps for Europe, Africa, the Middle East and Tehran). Where horizontal radiation is used, it should be adjusted for tilt and orientation angles (see: ***2.3 PV Module/Array Optimum Tilt and Orientation Angles***).

3. Electrical Energy Output Estimate for Grid-Tied PV System

3.1 Estimating PV System Electrical Energy Output

In designing a PV system, it is always necessary to make a proper estimate of the electrical energy output from the PV system, in order to take measures to conserve as much energy as possible and to be certain that the PV system meets the owner's expectations before carrying out the installation.

When the requirements for the installation of the PV system have been discussed, assessed, and agreed upon with the client (owner), and when all other measures – energy conservation, site survey, choice of type of PV module, inverter, including battery and controller in standalone PV systems, and selection of other PV system components, etc. – have been taken regarding the proper installation of the PV system, the sketch plans are developed. The sketch plans will be used for the initial estimation of the energy output of the PV system. The final estimation will be carried out when the PV system design has been completed. A big difference between the initial estimate and the final estimate can affect the terms of the contract.

When estimating PV system energy output (kWh), the estimation should be done in line with national codes and regulations, where this is required. The method of estimation may be regulated in order to avoid inconsistencies arising from different methods used in the estimation. For example, in the United Kingdom, a uniform calculation of PV system energy output has been regulated.

In general, the information required in order to estimate a PV system's annual electrical energy output can be as shown below:

- PV array area;
- PV module efficiency (PV module efficiency is often given in a PV module technical data sheet);
- Average solar radiation (kWh / m^2) for the location where the PV system is to be installed;
- PV module/array tilt and orientation angles;
- PV system performance factor.

Energy output (kWh) can then be estimated using a standard formula (universal estimation), as follows:

$$\begin{aligned}\text{Energy output (KWh)} \\ &= \text{PV array area} \times \text{PV module efficiency} \times \text{solar radiation (on horizontal plane)} \\ &\quad \times \text{factor for tilt and orientation angles} \times \text{PV system performance factor}\end{aligned}$$

Determining PV Array Peak Power:

When estimating a PV system's electrical energy output you need to establish the size of the PV array in peak wattage, at standard test conditions. The peak power of a PV module, at STC (at 1000 W / m^2 solar irradiance, at cell temperature of 25^0 C and 1.5 AM), can be obtained from the technical data sheet of the selected PV module or from the module manufacturer/supplier. The peak power of the PV array is the sum of the peak wattage (W) of individual PV modules making up the PV array and it can be determined as follows:

i) **For given or required yearly electrical energy consumption (unlimited by space and finance)**:

Divide the yearly required energy consumption by 365 to get daily energy consumption; divide the daily energy consumption (kWh / day) by average daily peak sun-hours, adjusted for tilt and orientation, to get the PV array peak power (kW) at ideal conditions (unadjusted for PV system performance factor); divide the peak wattage at ideal conditions by the PV system performance factor to obtain the required peak power of the PV system (see **Chapter 3**). So,

$$\text{PV array peak power} = \frac{\text{daily energy consumption}}{365 \times \text{ave peak sun hours adjusted for tilt and orientation} \times \text{system performance factor}}$$

Divide the above PV array peak power by selected PV module peak wattage to obtain the initial number of PV modules. So,

$$\text{Initial number of PV modules} = \frac{\text{initial PV array peak power}}{\text{selected PV module peak power}}$$

Multiply the initial number of PV modules by the peak power of the selected PV module to get the PV array peak power. So,

PV array peak power = initial number of PV modules × selected PV module peak power. The array peak power should not exceed array peak power obtained **from** limited available finance.

ii) **For limited available funds:**

Divide the amount of available finance by the cost per kilowatt peak of the PV system. This will give you the initial size, in kilo-watt peak, of the PV array for PV system. So,

$$\text{Initial PV array peak power} = \frac{\text{available funds}}{\text{cost per kilo} - \text{watt peak}}$$

Divide the above PV array peak power by the selected PV module's peak wattage to obtain the initial number of PV modules. So,

$$\text{Initial number of PV modules} = \frac{\text{initial PV array peak power}}{\text{selected PV module peak power}}$$

This PV array peak power (obtained for limited funds) should not be exceeded.

Multiply the initial number of PV modules by the peak power of the selected PV module to get the final PV array peak power. So,

$$\text{PV array peak power} = \text{initial number of PV modules} \times \text{selected PV module peak power}$$

iii) **For limited available space**:

For limited available space you will need to determine the area per kilowatt peak (m^2 / kW) of the selected PV module and divide the available space (m^2) by the PV module's area per kilowatt peak to obtain the initial PV array peak power. So,

$$\text{Initial PV array peak power} = \frac{\text{available space in square meters}}{\text{PV module area per kilo} - \text{watt peak}}$$

As before, divide the above PV array peak power by the selected PV module's peak wattage to obtain the initial number of PV modules. So,

$$\text{Initial number of PV modules} = \frac{\text{initial PV array peak power}}{\text{selected PV module peak power}}$$

Multiply the initial number of PV modules by the peak power of the selected PV module to get the final PV array peak power. So,

$$\text{PV array peak power} = \text{initial number of PV modules} \times \text{PV module peak power}$$

However, it is important to note that the initial number of PV modules may change, depending on PV module layout, which may either be landscape or portrait, and on the PV array and inverter compatibility (see section ***5.1.4 Deciding on PV module layout and number of PV modules** and **PV array/inverter compatibility***). The peak wattage, however, should not exceed the peak wattage obtained from available funds. If the peak power obtained for limited available space exceeds the peak wattage obtained **from** available funds, the peak wattage obtained for available funds should be used in the PV system design.

Determining PV module yield (PV module efficiency):

The PV module yield is the ratio of the PV module rated power, in kilo-watt, and the PV module area (m^2). This is the PV module efficiency factor (at standard test conditions), and it can be found in the module technical data sheet (see ***1.6 PV Cell/Module Peak Power and Efficiency***).

$$\text{PV module yield} = \frac{\text{PV module rated power (KW)}}{\text{PV module area (m2)}}$$

Finding global solar radiation (average peak sun-hours) or KWh / KW / day for the location:

The average peak sun-hours per day for the location can be obtained from global solar radiation maps, bar charts that can be found on some websites. (Note that average peak sun-hours is equivalent to global solar radiation.) Additionally, PV module yield for optimally tilted and oriented PV modules may be given as KWh / KW / day for certain locations (see ***2.5 Solar Resource***). The average sun-hours per day **can also be calculated manually from recorded radiation data**.

Determining PV module/array tilt and orientation angles:

As stated earlier, in desert areas, where the sky is often clear, and in lower latitudes the optimum tilt angle for the whole year is usually equal to the latitude angle, but in more northerly latitudes, due to cloudy, humid climatic conditions, the optimum tilt angle may be lower than the latitude angle (see ***2.1 Capturing of the Sun's Energy, 2.3 PV Module /Array Optimum Tilt and Orientation Angles, and 2.4 Example 2 – Radiation on a Tilted Surface***).

PV System Performance Factor:

The PV system performance factor is the product of the irradiance factor and the PV system efficiency factor:

i) Irradiance Factor:

This accounts for the fact that the PV module/array does not operate at the same irradiance as at standard test conditions (1000W / m^2). Manufacturers give power tolerances (usually from ±3% to ±5%) for the rated power of their PV modules. This is the PV module de-rate factor. It is necessary, therefore, to include this percentage loss due to module tolerance when determining PV array energy output.

The actual input power on a PV module/array at any location is equal to the operating irradiance (kW / m^2) multiplied by the effective area of PV module/array, and this power is often less than the input power obtained from irradiance at standard test conditions (see ***1.6 PV Cell/Module Peak Power and Efficiency***) multiplied by the effective area of the PV module/array. This is due to PV module tolerance, local weather conditions, dust accumulation on PV modules, light reflection from PV modules, and PV module temperature. The temperature under which the PV array operates may vary greatly, depending on the weather conditions where the PV system is located. When the temperature increases, it reduces the PV cell/module voltage as already stated (see ***1.3 PV Cell/Module Temperature***). The operating irradiance should include the temperature factor, i.e. it should be multiplied by the temperature factor. The standard applicable temperature factor for a crystalline PV module on energy output is 0.89. The ratio of the operating irradiance and the irradiance at standard test conditions is the irradiance factor. It is necessary, therefore, to apply the irradiance factor (which includes PV module tolerance) when determining PV array output. For a PV module rated at ±5%, if –5% is taken as the tolerance of the module, rated at 200W, the PV module rating is reduced by 5% and this means that – including irradiance reduction, due to weather conditions, dust accumulation on the PV module, PV module temperature, and light reflection from the PV module – the operating irradiance for the 200W PV module may be reduced to as low as perhaps 850W / m^2. This would reduce the PV module's power of 200W to 170W. Note that even for the best-designed PV system there will always be some soiling of PV modules, and therefore it is always necessary to apply a soiling factor on the PV array power output. A soiling factor of 0.98 to 0.95 (this can be much higher in desert areas) should be applied (see ***2.2 Other Factors that Determine PV Array Electrical Energy Output***) on a PV system's energy output. A total irradiance factor of 0.9 is reasonable. The irradiance factor should be applied to determine the electrical energy output under ideal operating conditions.

The irradiance factor is equal to 1 × product of factors mentioned above, or:

$$\text{Irradiance factor} = \frac{\text{operating irradiance } (\frac{\text{W}}{\text{m2}})}{\text{irradiance at standard test conditions } (\frac{\text{1000W}}{\text{m2}})}$$

ii) **PV system efficiency factor**:

The PV system efficiency factor is the product of the following factors:

- **PV module mismatch factor**:

Despite the fact that PV modules in a string must be of the same type, the same rating, the same tolerances, and have the same operating environment, there will still be some small power differences that will result in power losses. Therefore, it is reasonable to allow for power losses, due to the mismatching of PV modules and PV module wiring, of about 2%, giving a mismatching and wiring factor of about 0.98.

- **DC cable efficiency factor:**

The maximum permitted voltage drop of 3% for DC cable wiring (PV string cables and main DC cables to inverter) gives a DC cabling factor of 0.97 to be applied on PV array power output (check national codes and regulations).

- **Inverter efficiency factor:**

Inverter losses on both the DC side and the AC side of the inverter, should be taken into account when determining the power output of the PV system. Inverter conversion losses can reasonably be assessed as about 4% to 5% (consult selected inverter technical data for inverter efficiency under operating conditions); a factor of about 0.95 would be reasonable.

- **AC cable efficiency factor:**

Allow for an AC cable voltage drop of about 1% from inverter connection to utility supply, making a factor of about 0.99 (refer to national codes and regulations).

- **System availability factor:**

For grid-tied PV systems this can be 0.99. For standalone PV systems supplying critical loads this is 0.99, and for standalone PV systems supplying noncritical loads this is 0.95.

- **Charge controller efficiency factor and battery efficiency factor (for standby PV systems)**:

Where the PV system has one part supplying power to critical areas in the event of utility grid failure, or in standalone systems, factors for battery charge controller and battery/s (including DC cables from battery charge controller to battery) should be taken into account.

- **Shading factor:**

Note that partial shading is not part of the irradiance factor, and should be treated separately.

If there is partial shading of the PV array, the sun path (window) should be examined to find out the sources of partial shading (see ***1.9 Partial Shading***), e.g. by using a pathfinder or similar device to determine the shading factor. When the shading factor has been determined it should be included in the PV system performance factors when determining PV system energy output.

Determining estimated PV array output (Standard Estimation Method, UK):

The *Standard Estimation Method, UK* is carried out in accordance with section 3.7.2 of the ***Guide to the Installation of Photovoltaic Systems (2012).***

3.2 Example 3

A grid-tied PV system is expected to produce 2.5 kWh / kW per day at a given location. The PV array is at an optimum tilt angle of 30^0 and optimally oriented. PV modules will be cleaned regularly and there is no partial shading of PV modules at the location site. How much electrical energy, available to the consumer per day, will a 4.5 kW PV system, installed at the same location, yield when operating under similar conditions and with the PV array optimally oriented and tilted at optimum angle of 30^0? The PV system efficiency factor is expected to be equal to 0.86, comprising the following factors: PV module string cable 0.99, DC cables 0.98, inverter efficiency factor 0.90, AC cable 0.99, and system availability 0.99.

Solution:

The 4.5 kW PV system will be expected to produce four and a half times the amount of electrical energy per day than the amount produced previously. So,

Energy output (kWh) per day = 4.5 kW × 2.5 kWh / kW / day

= 11.25 kWh per day

Since the PV array is optimally tilted and oriented, the product of tilt and orientation factors is equal to 1. Since there is no partial shading, no shading factor is required. The electrical energy output obtained above is before taking into account the PV system efficiency factor of 0.86 (see ***3.1 Estimating PV System Electrical Energy Output***).

The above energy output (11.25 kWh) should now be multiplied by the PV system efficiency factor, in order to obtain the actual electrical energy output available to consumer. So,

Energy output (available to consumer) per day = 11.25 kWh × 0.86

= 9.67 kWh / day

3.3 Example 4

The consumer is expected to consume 5000 kWh per year from a grid tied PV system with PV array tilted at fixed angle of latitude 10^0 south, and oriented toward true north. Global irradiation at latitude tilt is 5.03 kWh / m^2 / day. The PV system performance factor is expected to be equal to 0.81 (made up of the following factors: soiling 0.98, temperature 0.89, PV module de-rating 0.97, PV module mismatch 0.99, DC cable 0.98, inverter efficiency factor 0.90, AC cable 0.99, and PV system availability 0.99). The PV module to be used is a 200W polycrystalline silicon module. What should be the size of the PV array? How many PV modules are required for the PV system?

Solution:

To obtain the amount of electricity (kWh) consumed per day, divide the amount of energy consumed annually by 365. So,

Electricity consumed per day = 5000 kWh ÷ 365 = 13.7 kWh / day

To obtain the PV array peak power, divide the electricity consumed per day by the average sun-hours per day (adjusted for tilt and orientation). So,

PV array power = 13.7 kWh / day ÷ 5.03 sun-hours / day) = 2.72 kW

Note that the factor for tilt and orientation is equal to 1 (factor at optimum tilt and orientation). In addition, the size (kW) of the PV array is for ideal conditions (STC), i.e. it does not take into account the performance of the PV system. To get the actual size of the PV array we should divide the peak wattage obtained above by the PV system performance factor. So,

PV array peak power (initial) = (2.72 ÷ 0.81) kW
= 3.358 kW

To calculate the number of PV modules, divide the initial PV array peak power by selected PV module peak power. So,

Number of PV modules = 3358 W ÷ 200W = 16.79. That means there are 17 PV modules (it is always better to round up than down). So,

Final PV array peak power = 200 W × 17 = 3.4 kW

4. Grid-Tied PV System

4.1 Grid-Tied PV System Main Components

4.1.1 Requirements of PV module for grid-tied PV system

The PV module should comply with internationally recognised standards. For example, the International Electro-technical Commission standard (IEC 61215 for crystalline PV modules and IEC 61646 for thin-film PV modules); the Committee of the European Community standard (CEC Approval Specification No.503); the European standard EN 50380; the UL (Underwriters Laboratories standard - USA), etc. PV modules complying with the above standards, or similar, are quality PV modules. PV modules from reputable manufacturers or suppliers can be reasonably assumed to be quality PV modules. Most international standards require that each solar module should have a technical specification data sheet. For example, the European standard requires that every PV module should have a technical data sheet showing, among other things, the following:

- Module peak power (W), including peak power tolerances
- Voltage at maximum power point (V_{MMP}) at 25^0 C
- Current at maximum power point (I_{MPP}) at 25^0 C
- Open-circuit voltage (V_{OC}) at 25^0 C
- Short-circuit current (I_{SC}) at 25^0 C
- PV module maximum reverse current
- Voltage temperature coefficient ($T_C - V_{OC}$)
- Current temperature coefficient ($T_C - I_{SC}$)
- Power temperature coefficient ($T_C - W$)

Note that the voltage coefficient is negative and the current coefficient is positive.

Other necessary information – such as the PV module reverse current, maximum system voltage, length, width, weight, protection against harsh weather conditions, glass, frame material – if not provided on the data sheet must be obtainable from the manufacturer. All of the above information is necessary for the proper design and installation of the PV system. For example:

a) Suitability for the system voltage:

The PV module should be class II equipment and should not be connected to the PV system/string with maximum PV array voltage ($V_{OC\ STC}$) or inverter maximum input voltage ($V_{DC\ MAX}$) – whichever is the lower – greater than the withstand voltage of the PV module. In other words, the PV string voltage is limited by the permitted maximum operating voltage at standard test conditions (maximum withstand voltage) of the individual PV module maximum withstand voltage) in a string and the maximum DC input voltage of the inverter, whichever is the lower. The selected PV module must have maximum withstand voltage ≥ the required system voltage. The maximum withstand voltage of a PV module can be obtained from that module's technical data sheet. If the maximum withstand voltage is not provided in the technical data sheet, the information may be obtained from the manufacturer.

b) Weather protection of PV module:

The PV module must be protected against the weather conditions in which it is going to operate, that is, against rain water, snow, hail storm, strong wind, and dust (see standards referred to above). This information is usually provided in the PV module's technical specification data; any of the information not provided can be obtained from the manufacturer.

4.1.2 Selection of PV module for grid-tied PV system

When choosing which PV module type to use one would expect that the owner would most likely want to use the most efficient PV module type available in the market in order to get more electricity from their PV system. Of the most available PV module types for supplying electricity in the buildings, the monocrystalline silicon PV module is the most efficient (see ***1.6 PV Cell/Module Peak Power and Efficiency***). However, it is important to note that if the owner has enough available space to accommodate a larger area of polycrystalline silicon PV modules, which can give the same amount of energy as that which can be obtained from the monocrystalline silicon modules, the owner should probably opt for the polycrystalline silicon module, since the polycrystalline silicon module is cheaper compared to the monocrystalline one. If the PV array is to be installed on the building's roof, upgrading of the roof may be required due to the additional weight from the polycrystalline silicon modules. Therefore, if the PV array is to be installed on the roof and there is enough space for the bigger polycrystalline modules and extra funds, which do not exceed the cost of monocrystalline PV modules, are available for upgrading the roof (to retain its strength and quality), the owner may opt for the polycrystalline modules. Where there is less direct sunshine, such as façades, and where PV module temperature is likely to be high, the owner should most probably choose the thin-film PV module, even though it has the lowest efficiency compared to monocrystalline and polycrystalline silicon modules, because it absorbs more diffuse radiation and is less susceptible to high module temperature. The information given in section ***4.1.1*** should also be considered. Additionally, the physical size and weight of the PV module need to be considered (see ***1.7 PV Module Physical Size and Weight***). Smaller PV modules are lighter and easier to handle compared to larger PV modules. For specific amount of energy required, smaller and more efficient PV modules are likely to occupy less of the limited available space than the larger and less efficient PV modules. How long the PV module will last is also very important. The majority of crystalline silicon solar modules commercially produced to date have a lifespan of 20 to 25 years. Warranties or guarantees should also be considered. Warranties given by some manufacturers of PV modules have 90% guaranteed PV module output for the first 10 years and 80% for the next 10 years. Note that warranties or guarantees from different manufacturers may be different. Warranties or guarantees must be carefully read and understood. For example, in replacing a defective PV module, conditions attached – relating, for example, to PV module maintenance, transport costs, the identity of whoever must carry out the actual installation, period required for replacement, etc. – will more likely be different for different manufacturers and they must be read and understood to avoid any inconvenience later.

4.1.3 Grid-tied PV inverters

Grid-tied inverters can be classified into two categories:

- Inverters that feed an isolated grid; if they have special parallel control, they can also feed a grid that is fed from other parallel sources. These inverters are called self-commutating fixed-frequency inverters.
- Inverters that will feed a grid only where the grid frequency is defined by another source, connected in parallel, and in the absence of the defined frequency (e.g. during grid black-out) they will not operate. These inverters are called line commuted fixed-frequency inverters.

These inverters can be subdivided further into single-string inverters, multi-string inverters, central inverters, and PV module inverters.

4.1.4 Grid-tied PV inverter types

PV string inverters:

There are two types of PV string inverters:

- **Single-string PV inverter:** This PV inverter is connected to only one PV string. It is for a small photovoltaic system made of PV modules connected in series. The power of a single-string PV inverter ranges from 0.7 kW to 2.5 kW. Several single PV strings each connected to a PV string inverter will be required for larger installations.

Advantages of the single-string PV inverter:

- Straightforward tracking of I-V curve maximum power point (MPP).
- Reduced DC cable lengths, since the PV string inverter can be placed nearer the PV array (PV main DC cables not required). This results in a reduced balance of system cost.
- PV combiner box (PV junction box) is not required. This results in a reduced balance of system cost.
- PV strings with different orientation and/or tilt can have a separate PV string inverter. Ongoing maintenance costs are lower (no fans or air filters).

Disadvantages of the single-string PV inverter:

- For larger PV systems, greater space is required for the installation of several single-string PV inverters, which results in increased costs.
- **Multi-string PV inverter:**

  On the DC side of the multi-string PV inverter, the connection is made up of several single PV strings, while on the grid (AC) side the connection is like a form of central PV inverter. The multi-string PV inverter is used mainly in PV installations for dwellings and commercial buildings.

Advantages of the multi-string PV inverter:

- One multi-string PV inverter can be used instead of several single-string PV inverters.
- A multi-string PV inverter tracks the I-V maximum power point (MPP) of each PV string individually.
- Provides choice to counter adverse effects of partial shading, orientation, or tilt of photovoltaic array.

Disadvantages of the multi-string PV inverter:

- Centralisation of the multi-string PV inverter results in greater PV DC cable (PV string cables) lengths and larger cross-sectional areas of DC cables.

Central PV inverter:

- On the DC side of the central PV inverter the PV string cables are connected together, in parallel, in the junction box (also known as the PV string junction box or PV string combiner box) and the main DC cables from the PV combiner box are connected to the central inverter.
- The central PV inverter ranges from about 4 kW to megawatts.

Advantages of the central PV inverter:

- High photovoltaic array output at high voltages.
- All PV string cables are connected in parallel in the PV junction box (combiner box).
- At lower voltages, of less than 120 VDC, the central inverter may be suitable where there is inevitable partial shading or where PV modules with large tolerances are used.

Disadvantages of the central inverter:

- The central inverter will not efficiently track the I-V maximum power point of the PV array that is made up of strings with an unequal number of PV modules.
- The centralisation of the inverter results in greater DC cable lengths and larger cross-sectional areas of DC cables.
- At lower voltages, of less than 120 VDC (because PV modules connected in series are fewer, while PV modules connected in parallel are several), the system DC current will be greater. This results in larger cross-sectional areas of DC cables.
- At high voltages the central PV inverter is not suitable where PV arrays are made of PV modules with high tolerances or where there is inevitable partial shading.

PV module inverter:

The PV module inverter is for one or two modules (for PV AC modules). Each PV system can have the PV module inverter factory fitted into the PV module junction box.

Advantages:

- Minimum loss of power due to partial shading – only one system is affected.
- PV modules with different tolerances can be used.
- Connection can be made direct to the utility grid (the PV module inverter is in the PV

junction box of the PV module and the AC cable from the PV module inverter can be connected direct to the grid).

- No PV string cables are required. That means reduced power losses in cables because AC cables will produce less voltage drop compared to the DC cables.

Disadvantages:

- Low power output level can be expensive for the system.
- The lifespan of the inverter, at present, is much less (10 to 15 years) than that of the PV module (20 to 25 years).

PV inverter charger (PV inverter battery charger):

PV inverter chargers, although they are generally used in standalone systems, may be used in grid-tied systems, where a standby PV system is required when the grid (AC) supply is switched off. The PV array assists the battery during daylight hours to provide the required supply of electricity, and during the night or on cloudy days when the supply is from the storage battery or batteries. The PV inverter battery charger provides protection against battery overload, low battery over-discharge, short-circuit, and reverse polarity, and keeps the battery within the rated battery temperature limits.

4.1.5 Requirements of grid-tied PV inverters

- The PV inverter technical data sheet should show the PV array peak power (to which the inverter will be connected). The inverter rated power should be less than the PV array/string peak power in order to make sure that the inverter will never be overloaded, especially if the PV array/string is optimally tilted and is oriented true south, for locations in the northern hemisphere or true north, for locations in the southern hemisphere (the inverter manufacturer should be consulted).
- The inverter must be capable of tracking the maximum power point (MPP) of the PV array.
- The maximum peak power tracking (MPPT) voltage of the inverter must be within the MPP voltage range of the PV array. This means that the PV array output voltage must never be greater than or lower than the inverter input voltage.
- The inverter must have a sine wave that is synchronous with the sine wave of the grid.
- The inverter must comply with national codes and regulations.
- The inverter must have high efficiency at full load and part load.
- The inverter must function well at both high and low ambient temperatures.
- The inverter must be suitable for the extreme weather conditions in which it may be required to operate.
- The inverter must operate automatically, that is, the inverter must be capable of breaking the circuit automatically when the utility power is switched off (planned or

unscheduled), to prevent islanding, a situation in which the inverter continues feeding the grid when the utility power is switched off. The inverter should remain switched off until the utility power is restored. The failure of the inverter to disconnect the circuit is undesirable with regard to public safety and that of the utility staff working on the grid, the quality of power supply, and likely damage to equipment in event of automatic or manual closure of the distribution system to a power island. Grid-tied inverters are sensitive to variations in grid voltage, frequency and impedance, and can only function within a specified range.* Therefore, before connecting the inverter to the grid, it is necessary to ascertain that the inverter complies with the relevant national codes and regulations. Discussion with the Utility during the planning stage is essential so that when the inverter is installed, the inverter functions as expected.

- The inverter should display photovoltaic array output and have fault indicators, and may have additional indicators such as for inverter temperature, power consumption, etc.
- The inverter must have a technical data sheet showing:
- Maximum photovoltaic array power ($P_{PV\ max}$)
    - DC nominal power ($P_{DC\ nominal}$)
    - Maximum power point tracking voltage, upper limit of V_{MPP} (V_{HIGHER})
    - Maximum power point tracking voltage, lower limit of V_{MPP} (V_{LOWER})
    - Maximum DC input voltage ($V_{DC\ max}$)
    - DC nominal current ($I_{DC\ NOMINAL}$)
    - Maximum DC current ($I_{DC\ MAX\ INPUT}$)
    - Nominal and maximum A/C current output
    - Nominal and maximum A/C power output
    - Efficiency at full load and part loads

***Note: Underwriters Laboratories 1741 (UL 1741) is a Standard for Inverters, Converters, Controllers, and Interconnection System Equipment for use With Distributed Energy Resources. The Institute of Electrical and Electronics Engineers 1547 (IEEE 1547) is a Standard for Interconnecting Distributed Resources with Electric Power Systems. The standard for anti-islanding protection in the United States of America has been UL 1741 harmonised with IEEE 1547, since 1999.**

4.1.6 Selection of grid-tied PV inverter

The type of grid-tie inverter to use in a PV system will depend on various factors, some of which are given below (see ***4.1.4 Grid-tied PV inverter types*** and ***4.1.5 Requirement of grid-tied PV inverters***). The inverter manufacturer, and national codes and regulations must also be consulted.

When choosing an inverter to use in a grid-tied PV system the following should be taken into consideration:

- Size of PV array peak power (W): The array peak power should be higher than the inverter power, usually about 105% to 110% or more (consult the inverter manufacturer). This takes into account the fact that while the PV array peak power is at standard test conditions the operating irradiance is usually below the irradiance at standard test conditions (1000W / m^2, at 25^0C, and 1.5AM). That means the inverter rated power versus PV array peak power will very much depend on the geographical location and weather conditions, PV array tilt, and orientation (see ***2.3 PV module/array Optimum Tilt and Orientation Angles***, and section ***5.1.4 PV module/array and PV inverter compatibility***). However, it is suggested that the inverter rated power should never be less than 90% of PV array peak power, especially if the PV array is optimally tilted and is oriented true south (for locations in the northern hemisphere) or true north (for locations in the southern hemisphere). This is to make sure that the inverter will never be overloaded. For most building roof-mounted grid-tied PV systems, the PV modules are not optimally tilted; likewise, for building façade-mounted grid-tied PV systems. PV inverters are usually designed to carry a little more than their rated power; always check with the PV inverter manufacturer. You should check the instructions from the inverter manufacturer and also national codes and regulations; for example, in the United Kingdom the inverter rated power ranges from 100% to 80% of PV array peak power typically.
- Whether PV strings have an equal number of PV modules or not: a PV array with PV strings of unequal number of PV modules should be connected to a multi-string inverter, while a PV array with an equal number of PV modules per PV string can be connected to either a central inverter or a multi-string inverter.
- Whether the PV array is free of shading or not.
- Whether all the PV strings have the same tilt and orientation angles or not.
- Whether the chosen PV module has small or large tolerances or not.
- The need for more than one inverter.
- PV module/string voltage variation due to temperature variations at the site (these must be within the inverter's MPPT voltage range and below the maximum DC voltage input of the inverter).
- Whether the inverter is suitable for the environment in which it is required to operate.
- Inverter efficiency at full and part load.
- Cost of the inverter, including transport cost, should be affordable.
- Warranties and lifespan of the inverter.

4.1.7 PV module/array and inverter compatibility

In a grid-tied PV system, PV modules are usually connected in series, but they are connected in series and parallel where: there is partial shading to mitigate power output losses due to partial shading; the PV modules have high tolerances or the electrical characteristics of PV modules in a PV string differ, in order to reduce the effect of mismatching; and where there is necessity to maintain output voltage so that the PV array voltage remains lower than the input voltage of the inverter (See ***4.1.4 Grid-tied PV inverter types***).

The output voltage and output current of the PV module/array must not exceed the maximum DC input voltage and the maximum DC input current of the inverter, respectively. If the maximum DC input voltage and the maximum DC input current of the inverter are exceeded the **inverter can be damaged**. Additionally, the output voltage and the output current of the PV array must not be below the lower limit DC input voltage and nominal DC input current of the inverter. If the output voltage of the PV array falls below the lower limit of the inverter DC input voltage, the inverter will **not track the maximum power point (MPP) of the PV array**. Therefore:

- Open-circuit voltage (V_{OC}) of the PV module/array, at lowest temperature at the site, must not be greater than the maximum DC input voltage ($V_{DC\ MAX\ INPUT}$), of the inverter.
- PV module/array current at maximum power point (I_{MPP}) must not exceed the maximum DC input current ($I_{DC\ MAX\ INPUT}$) of the inverter.
- The maximum power point (V_{MPP}) voltage range of the PV array must be within the maximum peak power tracking (MPPT) voltage range of the inverter. This means that the V_{MPP} of the PV module/array, at the lowest temperature at the site, must not exceed the higher limit of the MPPT voltage of the inverter (V_{HIGHER}), and that the V_{MPP} of the PV module/array, at the highest temperature at the site, must not be lower than the lower limit of the MPPT voltage of the inverter (V_{LOWER}).
- The maximum number of PV modules in a PV string should be such that the PV string voltage, at maximum power point (V_{MPP}) and at lowest temperature of the site, does not exceed the higher limit of the maximum peak power tracking voltage of the inverter (V_{HIGHER}). So,

$$\text{Maximum number of PV modules} \times V_{MPP\ AT\ LOWEST\ SITE\ TEMPERATURE} \leq V_{HIGHER}$$

- The minimum number of PV modules in a PV string should be such that the PV string voltage, at maximum power point (V_{MPP}) and at highest temperature of the site, does not fall below the lower limit of the maximum peak power tracking voltage of the inverter (V_{LOWER}). So,

$$\text{Minimum number of PV modules per PV string} \times V_{MPP\ AT\ HIGHEST\ SITE\ TEMPERATURE} \geq V_{LOWER}$$

- The number of PV modules in a PV string should be such that the PV string open-circuit voltage (V_{OC}), at the lowest temperature of the site, will not be greater than the inverter maximum DC voltage input ($V_{DC\ MAX\ INPUT}$). So,

$$\text{Maximum number of PV modules per PV string} \times V_{OC\ AT\ LOWEST\ SITE\ TEMPERATURE} \leq V_{DC\ MAX\ INPUT}$$

- The PV array current at maximum power point (I_{MPP}) should not exceed the inverter maximum DC current input ($I_{DC\ MAX\ INPUT}$). So,

$$\text{PV array current at maximum power point (sum of PV string currents)} \leq I_{DC\ MAX\ INPUT}$$

The peak power of the PV module/array must be equal to, or slightly greater than the inverter power; in fact, where the PV array is optimally tilted, it should never be more than 110% of the inverter rated power (see ***4.1.6 Selection of grid-tied PV Inverter***).

4.1.8. Inverter Location

The inverter should be installed as near as possible to the PV array or PV array junction box so that the PV string cables and main DC cables are kept as short as possible. The inverter should be installed where it will not create disturbing noise (noise pollution) to the people living in the building. The inverter will operate efficiently when it is operating within its rated temperature. Excessive inverter temperature leads to reduced inverter efficiency. Additionally, the inverter should be protected against dust and humidity. In fact, the inverter location must comply with the instructions of the inverter manufacturer.

4.2 Other Components for Grid-tied PV Systems

4.2.1 PV module junction box

The PV module junction box is factory fitted and houses the PV module reverse current fuse, bypass diodes, and connector. The connector provides interconnection between the PV modules and connection between the PV module and the load. The PV module junction box is usually sealed by the manufacturer unless it is specified by the manufacturer as field serviceable.

PV cells as well are provided with bypass diodes. The bypass diode allows the electrical current to bypass a shaded or defective PV cell/module, thereby preventing hotspots (in case of damaged PV cell/s) or further damage to the PV cell/module or possible fire.

4.2.2 PV array junction box

The PV array junction box (also known as PV array combiner box), apart from providing a parallel connection of PV strings and connection to main DC cables, houses blocking diodes, surge diverters (if required), and PV string fuses. A disconnection switch/isolator, for isolating the PV array junction box from the load, should always be installed between the PV array and the inverter and positioned closer to the inverter. Some inverters have this disconnection switch as part of the inverter. Where the junction is situated far from the inverter (say, 10 metres or more) another disconnection switch or isolator may be installed close to the junction box or housed in the junction box. Usually, inverters are installed close to the junction box to reduce voltage drop. In larger PV systems, blocking diodes may be provided for each PV string (see **Figure 6, Figure 10** and **Figure 11**). As explained earlier (see ***1.1 PV Cell/Module Technology***), bypass diodes will allow the current to bypass shaded or defective PV cells, and the blocking diode will prevent the flow of electrical current from healthy PV strings into PV strings with shaded, failed, or defective PV modules and in standalone PV systems will prevent current from the battery to the PV array. **The PV string fuse** may also be used as a means of isolating the PV string during maintenance or testing. A PV array junction box must have a voltage rating calculated in accordance with the PV module manufacturer's instructions, and national codes and regulations. The current rating of the PV array junction box should be equal to or greater than the expected current of the circuit to which it is connected. The voltage and current ratings of PV array junction box are derived from the PV array V_{OC} at STC and the expected I_{SC} at STC, respectively, (check PV module manufacturer's instructions, and national codes and regulations).

A PV array junction box must be suitable for the environment in which it is supposed to work and should meet international standards and national codes and regulations.

4.2.3 PV module supporting structure

The PV module supporting structure should be compatible with the PV module frame and the building's roof structure (where PV modules are installed on the roof). In fact, it is advisable that the PV module supporting structure be ordered from the supplier/manufacturer of the PV modules for the PV system. In addition, if the supporting structure is purpose made, it should be made of such materials that are compatible with the PV module frame and the building's roof structure materials to which the supporting structure will come into contact. Brackets and bolts (all fixtures and fasteners) for the supporting structure should be of such materials, for example, stainless steel or galvanised steel, which will not cause corrosion, and all penetrations on the building roof should be water-tight or sealed. When PV modules are installed on the building's roof, care should be taken to ensure that the roof is not impaired or compromised. In other words, the integrity of the roof must be maintained. The weight of the PV modules plus the weight of the PV supporting structure (including wind force) should be such that the weight can be sustained by the roof of the building. The PV module structure should be capable of withstanding lateral and vertical forces that may occur in the site location. The calculation of the weight exerted on the roof by the PV array, to find out whether the weight is sustainable, should be done in accordance with the national codes and regulations. It may be necessary to check with the roof installer before carrying out the PV array installation or to request the services of a structural engineer. The installation of the PV array's supporting structure, whether on the building's roof or on the ground, should be carried out in accordance with the instruction from the manufacturer of the PV module supporting structure and in compliance with national codes and regulations. PV array supporting structures that can be adjusted for optimum tilt angle, when installed on a building's roof, should comply not only with national regulations but also they may require compliance with local council rules and regulations.

4.2.4 DC cables

PV cables must meet some internationally recognised standards and should comply with national codes and regulations. The DC cables must be suitable for the environment in which they are supposed to work. The DC cables must be capable of operating at extreme temperatures, e.g. -40^{0}C to 120^{0}C, and be double insulated (class II), UV and flame resistant, low smoke, halogen free, and water protected. Preferably, the DC cables must be light, flexible (with stranded conductors), and therefore easy to work with.

DC cables should have voltage rating not less than the PV array open circuit voltage multiplied by a temperature correction factor, and each cable should have a current rating greater than or equal to the expected current of the circuit (check manufacturer's instructions, and national codes and regulations).

4.2.5 Fuses, switches (isolators, connectors, circuit breakers, switch fuses)

All fuses, switches, isolators and connectors should be weatherproof or suitable for the environment in which they are supposed to work. Fuses and switches should be suitable for operation in DC circuits. It should also be noted that, on Utility grid connected PV systems, the DC circuit may only be opened or switched off when the AC side (of the inverter) has been switched off. Switching off DC side of the inverter when the AC side of the inverter is switched on can damage the inverter (check national codes and regulations and instructions of the inverter manufacturer). String fuses are preferred to blocking diodes (check national codes and regulations). String fuses should be type g PV and for operation at high temperatures. PV string fuse may be used for isolating the PV string from the load (the rest of the system).

The fuses, switches, isolators and connectors should be rated at no less than the maximum PV string (array) open circuit voltage at standard test conditions (U_{OC} at STC multiplied by a temperature correction factor), and have a current rating greater than or equal to expected current (I_{SC}) value (check manufacturer's instructions, and national codes and regulations).

4.2.6 Meters

Meters are usually the property of the utility and their installation is usually carried out by the utility. The metering in a grid-tied system can be divided into the following three types:

(i) Offset metering: The offset meter is not connected to the incoming utility grid but to the consumer unity (distribution board), usually via a utility switch. The electricity from the PV system is used in the building offsetting the electricity from the grid that would otherwise be consumed. The offset meter measures the PV system's electrical output.

(ii) Import-export metering: This type of metering can have two main meters – one for imported electricity (electricity consumed) and the other for exported electricity or can have one meter measuring both imported and exported electricity. This type of metering may have a special meter for measuring the electricity produced by the PV array.

(iii) Export metering: The export meter is connected to the incoming grid, before the main utility meter. In this type of metering all the electricity from the PV array is exported to the utility grid.

5. Designing a Grid-Tied PV System

5.1 The Steps for a Grid-Tied PV System Design

- Initial discussion with client;
- Site inspection;
- Preliminary design and sketch plans, and initial estimate of energy output (KWh) of the PV system if required;
- Design and final estimate of energy output (KWh) of the PV system.

5.1.1 Initial discussion with client (owner)

The discussion with the client on the installation of the grid-tied PV system should focus on finding out the client's requirements.

The designer or installer must find out the size (kW) or the amount of electrical energy the client expects from the PV system. The owner may be aiming at installing a PV system that will give him/her energy output that is enough for their monthly energy consumption. The client may want to install the PV system to reduce their electricity bill from the utility, or to sell part or all the electrical energy from the PV system to the utility. The owner should be made aware of the limiting factors. For example, if the available area for the PV system is limited, the peak wattage of the PV system may be less than the finance allows, especially if the PV array is to be mounted on a building's roof. On the other hand, if the available area for the PV array is adequate, the anticipated size (kW) of the PV system may be limited by the available finance.

The client should be made aware that he/she (the client) will need to find ways of conserving as much electricity as possible. For example, the client can do this by installing efficient lights and other electrical appliances that consume less electricity, and by making sure that lights are switched on only when needed. The designer/installer should explain to the client about available government incentives, if any, on renewable energy installations. That means that the designer/installer should explain about offset metering, and whether tariffs are available, explaining also how they operate and how much is being offered for electrical energy that is exported to the utility grid. The incentives may also be in the form of reduced tax on PV materials, of non-taxable PV materials, or of cheap loans for the installation of PV systems, etc.

The client may be prepared to spend a fixed amount on the PV system, which limits the peak wattage of the PV array. The designer should establish whether the client (owner) has adequate finance to meet the cost of the required PV system, including maintenance costs. The client may decide which type of PV module he/she would like for his/her PV system. The client's choice of PV module type could be based on which PV module is most efficient, or on aesthetics. On the other hand, the client may want a cheaper PV module and a cheaper inverter in order to reduce the initial cost of the PV system. The designer/installer should explain clearly to the client the advantages and disadvantages of most available PV module types (see: ***4.1.2 Selection of PV***

module for grid-tied PV system). Regardless of what type of inverter the client wants, the final decision on the type of inverter to be used will be made at a later stage in the design.

For example, while the monocrystalline PV module is the most efficient among most of the commercially available PV modules, it is also the most expensive. If the client has adequate finance and the available space is limited they may choose the monocrystalline in order to increase the size (kW) of the PV system. The less expensive polycrystalline PV module can produce the same amount of required electricity, but the PV array composed of polycrystalline PV modules will require more space (see ***1.6.1 Example 1***).This also refers to thin-film PV modules: while they are much cheaper, they are also less efficient (see ***1.6 PV Cell/Module Peak Power and Efficiency***). The designer/installer should also explain to the client the advantages of having a multi-string inverter instead of a central inverter. For example, while a multi-string inverter is more expensive than a central inverter, the former is more efficient where PV modules have higher tolerances and where the PV strings have a different number of PV modules; in addition, the multi-string inverter does not require a junction box and is easy to install (see ***4.1.3 Grid-tied PV inverter types***). The designer/installer should also explain to the client, clearly and in simple terms, how the PV system works. He/she should explain about the possible lifespan of the system's main components and the required maintenance to prolong their lifespan. He/she should emphasise the importance of maintaining the PV system, in accordance with the instructions given, so that it can produce as much electricity as expected. The maintenance required is mainly regular cleaning of the PV modules and removal of anything that can cause any shading of the PV modules. (In standalone PV systems, maintenance may include topping up of the battery.) In addition, constant monitoring of the PV system's power output, which can be read from the PV array meter, may be required, to check that the PV system is performing properly or as expected. The client must also be made aware that the weather conditions, dust, and dirt can reduce the electricity output from the PV system. Even though the initial cost of installing the PV system at present may seem to be too high because of the high cost of the system's main components, the advantages of installing it can outweigh the cost. When a PV system is properly designed, installed, and maintained, it is capable of paying itself off much earlier than its lifespan. The designer/installer can make an estimate of the payback period of the PV system, and explain this to the client. Most PV modules have a guarantee of 20 to 25 year lifespan, and some batteries (for standby or standalone PV systems) and inverters for PV systems have 10 to 15 years lifespan.

Once the designer/installer has explained clearly to the client/owner of the PV system how it will work, and the client fully understands what they should expect from the PV system, the client will be able to make a better and informed decision as to whether he/she can install the PV system or not. Once the PV system is installed the client will be able to tell whether it is performing to expectation or not.

5.1.2 Site visit

The designer/installer should visit the site to find out whether it is suitable for the installation of the PV system, and whether the available area will be adequate for the required size (kW) of the PV array.

If the PV array is to be mounted on the building's roof, the designer/installer should verify whether the available building roof area is adequate for the size of the PV system required. He/she should check the roof pitch and orientation. The PV array should be mounted on the side of the roof oriented toward true south for a location in the northern hemisphere, and oriented toward true north for a location in the southern hemisphere (note that, for standalone PV systems, PV modules must always be optimally tilted and oriented). However, for grid-tied PV systems, a small deviation from true south or true north and from optimal tilt angle can be acceptable, as this has very little effect on the output of the system (see ***2.3 PV Module/Array Optimum Tilt and Orientation Angles***). If the PV modules are on a pitched building roof, check whether the roof pitch will be adequate for PV module (array) cleaning. In grid-tied PV systems, PV modules are usually laid flat on the roof structure. A minimum tilt angle (pitch) of 12^0 is said to be adequate for PV module cleaning.

If the PV array will be installed on a flat roof or on the ground, the designer should verify whether the available area is suitable and adequate for the installation of the PV system. The available area for the PV array should include a row spacing area (see ***1.9 Partial Shading***).

The designer should decide whether the PV modules will be laid out in portrait or landscape (this may change later) and make an initial estimate of the size (kW) of the PV array. The site should be checked to make sure it is free of any shading or potential shading. This can be done by using a solar path finder. Any form of shading should be avoided or removed if possible.

The designer or installer should find solar radiation (kWh / m^2 / day) and wind speed for the site/location. The accuracy in the determination of the solar resource (solar radiation) that will be used will greatly determine the output of the PV system. Incorrect (under- or over-valued) solar radiation (kWh / m^2 / day) can result in the PV system failing to work to expectation (see ***2.5 Solar Resource***). He/she must find out the minimum and maximum temperatures at the site location.

The designer/installer should find a suitable space where the inverter can be installed (see ***4.1.8 Inverter location***). He/she should decide on cable routes making sure that the DC cable lengths are as short as possible so that the voltage drop to the inverter does not exceed 3% (check national codes and regulations), and calculate the cable lengths.

5.1.3 Preliminary design and estimation of the PV system energy output

Decide on the size of the PV array and on energy conservation. Confirm the position of the inverter. Make sketch plans of the PV system layout, which should indicate length of PV string cables and main DC cables, and decide on cable installation method.

Make the initial estimate of the PV system electrical energy output. This will be carried out as indicated earlier (see ***3.1 Estimating PV System Electrical Energy Output***).

5.1.4 Design stage

Determining number of PV modules and PV array peak power:

The number of PV modules and the PV array peak power can be determined as shown earlier (see ***3.1 Estimating PV System Electrical Energy Output***).

Deciding on PV module layout and number of PV modules:

Determine the number of PV modules, laid in portrait or landscape, which will fit into the available space. Decide on how many rows of PV modules, and the number of PV modules per row, for each layout format. The calculation for PV module layout can be carried out as shown in ***5.3 Example 6*** or ***5.4 Example*** *7*. For PV arrays installed on building roofs, steps should be taken to make sure that the PV modules will fit into the available roof space, leaving enough walking space for PV system maintenance.

Compare the initial number of PV modules (see ***3.1 Estimating PV System Electrical Energy Output***) with the number of PV modules obtained by portrait or landscape layout format to see what number of PV modules can be adopted. Select the number of PV modules to be used; these should be laid out in their respective PV module layout format. The initial number of PV modules can be selected if they fit into either the portrait or landscape format, and if selected they should be laid out in the corresponding PV module layout format. When PV modules are laid on the ground or on flat roofs the landscape format is likely to be adopted, in order to reduce the distance between the rows of PV modules, while making sure that the front row of PV modules does not shade the one just behind. However, the total number of PV modules and the number of PV modules per PV string will be decided at a later stage (see below: ***4.1.7 PV module/array and PV inverter compatibility***).

Determining PV module voltages at minimum and maximum temperatures at the installation site:

The PV module voltage at extreme temperature is equal to the PV module voltage (STC) plus change in voltage due to temperature change or variation. So,

$V_{(AT\ EXTREM\ TEMPERATURE)} = V_{STC} + (T_C - T_{STC}) \times (T_{COEFF})$, where: V_{STC} is the voltage (V_{OC} or V_{MPP}) at STC; T_C is the operating temperature; T_{STC} is the temperature at STC; T_{COEFF} is the temperature coefficient. Therefore, the PV module voltage at extreme temperatures is calculated as follows:

V_{MPP} at lowest site temperature = V_{MPP} at STC + (- difference in temperature × - voltage temperature coefficient)

V_{MPP} at highest site temperature = V_{MPP} at STC + (difference in temperature × - voltage temperature coefficient)

V_{OC} at lowest site temperature = V_{OC} at STC + (- difference in temperature × - voltage temperature coefficient)

Remember that the voltage coefficient is negative.

Determining size (kW) of inverter and inverter selection:

Determine the inverter size in kilowatts. This is the rated inverter output power (see ***4.1.5 Requirements of grid-tied PV inverters*** and ***4.1.6 Selection of grid-tied inverter for PV system***). Obtain the inverter technical data sheet.

Determining the maximum and minimum number of PV modules per PV module string (PV string):

The number of PV modules per PV string will depend on the PV system's required voltage and the inverter maximum peak power tracker (MPPT) voltage range.

Once you have the inverter technical data sheet, using the inverter MPPT voltage range, inverter DC maximum voltage input and the PV module V_{MPP} and V_{OC} (both at extreme PV module temperatures at the site), you can calculate number of PV modules per string as follows:

i) Maximum number of PV modules per PV string ≤ inverter higher limit of maximum power point tracking voltage (V_{HIGHER}) ÷ PV module maximum power point voltage, at lowest PV module temperature at the site ($V_{MPP\ AT\ LOWEST\ PV\ MODULE\ TEMPERATURE}$)

ii) Minimum number of PV modules per PV string ≥ inverter lower limit of maximum power point tracking voltage (V_{LOWER}) ÷ PV module maximum power point voltage at the highest PV module temperature at the site ($V_{MPP\ AT\ HIGHEST\ PV\ MODULE\ TEMPERATURE}$)

iii) Maximum number of PV modules per PV strings ≤ inverter maximum DC input voltage ($V_{DC\ MAX}$) ÷ PV module open-circuit voltage at the lowest PV module temperature at the site ($V_{OC\ AT\ LOWEST\ PV\ MODULE\ TEMPERATURE}$)

It is important to remember that the PV module string voltage must never exceed the PV module maximum rated voltage (given by the PV module manufacturer).

Having calculated the maximum and minimum number of PV modules per PV string you should decide on the number of PV modules per PV string to use and determine the number of PV strings that can be formed from the selected number of PV modules, which should be laid out in the corresponding PV module layout format, i.e. landscape or portrait format. The selected number of PV modules can be the initial number of PV modules. So,

Number of PV strings = selected number of PV modules ÷ selected number of PV modules per PV string

PV strings with greater number of PV modules are preferred, because they have higher voltage and therefore less voltage drop; this means that they result in reduced DC cable sizes and higher system efficiency. The selected inverter type will also have an influence on the decision concerning the number of PV modules for the PV array. For example, a central inverter will require that the PV array should have PV strings of equal number of PV modules in order to enable the inverter to track the PV array maximum power point efficiently, while a multi-string inverter will efficiently track the maximum power point of PV array with strings of unequal number of PV modules per PV string. For example, if you have a PV array with two PV strings of eight PV modules per PV string and two PV strings of 10 PV modules per PV string, the arrangement could be as follows:

- You could have two PV arrays, one PV array with one central inverter connected to two PV strings of eight PV modules, and the other PV array with a separate central inverter connected to two PV strings of 10 PV modules;
- Alternatively you could have one PV array of two PV strings of eight modules per string, and two PV strings of 10 PV modules per PV string connected to one multi-string inverter.

Remember that, in a PV string, PV modules must have the same tilt and orientation angles. PV module strings connected to a central inverter must have similar tilt and orientation angles to enable the inverter to track the maximum power point (MPP) of the PV array. The multi-string inverter is capable of tracking the MPP of a PV array made of PV module strings of different tilt and orientation (see ***4.1.4 Grid-tied PV inverter types***). Therefore, it is necessary that you choose the number of PV modules (for the PV array) that will be suitable for the selected inverter type.

Once the decision on the number of PV modules per PV string has been made, make sure that the voltage at maximum power point of the selected PV string is within the selected inverter's MPPT voltage range and that the PV string open-circuit voltage, at lowest PV cell temperature at the site, is below the maximum DC input voltage of the inverter.

PV module/array and PV inverter compatibility:

Make sure the PV module/array and the PV inverter are compatible as follows:

i) The maximum number of modules in a PV string should be such that the PV string voltage, at maximum power point (V_{MPP}) and at the lowest temperature of the site, does not exceed the higher limit of the maximum power point tracking voltage of the inverter (V_{HIGHER}). So,

Maximum number of PV modules $\times V_{MPP\ AT\ LOWEST\ SITE\ TEMPERATURE} \leq V_{HIGHER}$

ii) The minimum number of modules in a PV string should be such that the string voltage, at maximum power point (V_{MPP}) and at the highest temperature of the site, does not fall below the lower limit of the maximum power point tracking voltage of the inverter (V_{LOWER}). So,

Minimum number of PV modules $\times V_{MPP\ AT\ HIGHEST\ SITE\ TEMPERATURE} \geq V_{LOWER}$

iii) The number of modules in a PV string should be such that the PV string open-circuit voltage (V_{OC}), at the lowest temperature of the site, will not be greater than the inverter DC maximum voltage input ($V_{DC\ MAX\ INPUT}$). So,

Maximum number of PV modules $\times V_{OC\ AT\ LOWEST\ SITE\ TEMPERATURE} \leq V_{DC\ MAX\ INPUT}$

iv) The PV array current at maximum power point (I_{MPP}) should not exceed the inverter DC maximum input current ($I_{DC\ MAX\ INPUT}$). So,

PV array current at maximum power point (sum of PV string currents) $\leq I_{DC\ MAX\ INPUT}$

Note that the PV string voltage for PV arrays installed on buildings should never exceed the maximum permissible on the building (consult national codes and regulations).

Selection of cable types, and calculation of cable and other DC component sizes:

The DC cables must be suitable for PV systems, and cable cross-section areas should be calculated in accordance with national codes and regulations; this applies also to the selection of other DC components (see ***4.2 Other Components for Grid-tied PV Systems***).

Selection of PV module supporting structure and accessories:

PV module supporting structure and accessories should preferably be obtained from the manufacturer of the PV modules. The supporting structure and accessories should be compatible with the PV module frame and building roof material (see ***4.2.3 PV module supporting structure***).

Final estimate of PV system electrical energy output:

Calculation of estimated electrical energy output is as described in ***3.1 Estimating PV System Electrical Energy Output***.

5.2 Example 5

The owner wants at least 7000 kWh, per year, from his grid-tied PV system. The PV array is to be installed on the building roof, and the roof space for the PV array is six metres long and four metres wide. There is no limit to the availability of funds. Calculate the kilowatt peak (kW) of the PV array. Use the polycrystalline PV module rated at 200 W. There is no partial shading of the array. The building roof pitch is 35^0 (fixed optimum tilt angle year-round for the location = 35^0) and the roof is oriented due true south. That means that the tilt and orientation factor is equal to one.

The number of sun-hours per day is 4.0. The performance factor is as follows:

- There are no interruptions to the operation of the PV system
- Inverter efficiency is 96.5%
- PV module tolerance (de-rating factor) is ±0.95
- Module temperature de-rating factor (due to temperature losses) is 0.89
- Module mismatch de-rating factor (due to temperature losses) is 0.99
- DC cables de-rating factor (due to cable losses) is 0.98
- AC cables de-rating factor is (due to cable losses) is 0.99
- PV system availability is 0.99
- Module soiling de-rating factor (due to shading of PV cells by dust and dirt) is 0.97

Solution:

To find the owner's average daily consumption, divide the annual electrical energy required (7000 kWh) by the number of days in a year (365), i.e. 7000 kWh ÷ 365 = 19.178 kWh per day.

To find the size of the PV array under ideal conditions, divide the daily consumption by the number of daily sun-hours (5.0 sun-hours per day), adjusted for tilt and orientation, and by PV module de-rate factor (factor for tilt and orientation = 1). So,

PV array peak power = 19.178 kWh / day ÷ (4 sun-hours / day × 1 × 0.95) = 5.05 kW (approximately 5 kW). This PV array peak power is for ideal conditions.

In order to determine the size (kW) of the PV array under actual operating conditions, with de-rating factors as listed above, divide the PV array kilowatt peak (kW), obtained under ideal conditions, by the PV system performance factor (η).

The performance factor (η) of the PV system is equal to the product of the de-rating factors. So,

$\eta = 0.965 \times 0.89 \times 0.99 \times 0.98 \times 0.99 \times 0.97 = 0.80$ (approximately). Therefore,

PV array peak power (KW) = 5.05 kW ÷ 0.80 = 6.313 kW.

Number of PV modules = 6313W ÷ 200W = 31.57. This means that the PV system will require 32 PV modules (approximately). Therefore, the actual PV array peak power required is equal to 32 PV modules multiplied by 200W. So,

PV array peak power = 32 × 200W = 6400W

5.3 Example 6

Prepare the initial PV system's estimated power output (no disclaimer required), design a grid-tied solar electric system, and draw the electrical diagram showing the main components of the PV system for a house in Norfolk, UK.

From discussion with the client:

- The PV array is to be installed flat on the building's roof.
- No standby supply is required.
- The PV supply is to be based on the government's feed-in-tariff scheme: an import meter for registering electricity imported from the utility grid is required; a PV arrray meter for indicating the electricity produced by the PV array is also required; under the scheme, the PV system's AC power output should be such that the PV system's AC current does not exceed 16A per phase.
- There are no limitations on finance.
- Either monocrystalline or polycrystalline PV modules are to be used.
- Either a central inverter or a multi-string inverter can be used.

From site inspection:

- Site location is Norwich, Norfolk – latitude 52^0
- The available roof area is 8m x 4m
- The roof is oriented due south
- The building roof pitch is 35^0
- The roof is free of any shading
- Inverter to located in the loft
- Zone 12
- Postcode NR
- KK is 961 kWh/kW
- Temperaure range at the site is from -10^0C to $+40^0$C
- Expected PV module temperature range is from -10^0C to $+70^0$C

***PV module choice**:*

The type of photovoltaic module to be used is a KT200 multi-crystalline PV module with a power rating of 200W peak power (200W), and with a power rating tolerance of +10% to –5%. It has an efficiency of 16%; it is double insulated (class II); it is protected against severe environmental conditions; it has an anodised aluminium frame; it has a 25 year guarantee. The technical data sheet for this PV module is shown below.

PV module technical data (at STC):

• Peak power	200W (W)
• Tolerance	+10% / –5%
• Maximum power point voltage (V_{MPP})	26.3V
• Maximum power point current (I_{MPP})	7.61A
• Open-circuit voltage (V_{OC})	32.9V
• Short-circuit current (I_{SC})	8.21A
• Reverse current	15A

Thermal parameters:

• I_{SC} temperature coefficient	(3.18×10^{2}) A / ^{0}C
• V_{OC} temperature coefficient	(-1.23×10^{-1}) V / ^{0}C
• Normal operating cell temperature	47^0C

Mechanical parameters:

• Length	1.425m
• Width	0.99m
• Depth	0.36m

- Weight 18.5kg

Initial PV System Energy Output Estimate:

Determining the number of PV modules and PV array peak power (W):

From the electrical specifications of the PV module, the length of the PV module is 1.425m, the width is 0.99m, and the peak power is 200W.

Therefore, the area of the PV module per kilowatt peak = (1.425m x 0.990m) / 0.2 kW, i.e. 7.05m^2 / kW.

The initial PV array peak power is equal to the available roof area (m^2) divided by the area of the selected PV module per kilowatt peak (m^2 / kW). (The available roof area is assessed to be 80% of the roof area.) So,

PV array peak power = (8 x 4) m^2 ÷ 7.05m^2 / 1 kW = 4.539 kW

Number of PV modules is the PV array peak power (4539W) divided by PV module peak power (200W). So,

Number of PV modules = 4539 W ÷ 200 W = 22.69

This gives an initial number of 22 PV modules (rounded down).

Since the PV array peak power is equal to the PV module peak power multiplied by the number of modules, so,

PV array peak power = 200W x 22 = 4400W.

The 22 PV modules are not acceptable. The tariff scheme requires that the kW should not exceed 4kW, so that the PV inverter phase current does not exceed 16A. A PV array with a peak power of 4000W will fit into the available roof area and will have the following number of PV modules:

Number of PV modules = 4000W ÷ 200W = 20.

The inverter power (W) at the inverter terminals should be between 9.5 and 0.9 of PV array peak power, and should never be less than 0.9 where the PV array is optimally tilted and oriented (***4.1.6 Selection of grid-tied PV inverter***). This means that the inverter rated power is between: 4000W × 0.95 = 3800 and 4000 × 0.9 = 3600.

If the selected PV inverter, with the rated output power between 380W and 3600W, is a 3600W inverter, then:

PV inverter output current = 3600W ÷ 230 = 15.65A. This is acceptable because the PV inverter output current per phase is below 16A, as required by the tariff scheme.

However, in this initial estimation of the PV system energy output the 20 PV modules are reduced to 18 modules. These 18 PV modules will give a required peak wattage of 3600W (18 × 200W).

PV inverter power (output terminals) should be between: 3600W × 0.95 = 3420 and 3600 × 0.9 = 3240W.

If the selected PV inverter, with the rated output power between 3420W and 3240W, is a 3400 W PV inverter, then:

PV inverter output current = 3400W ÷ 230 = 14.78A. This is acceptable because the PV inverter output current per phase is below 16A, as required by the tariff scheme. Therefore, the 3600W PV array is acceptable.

If an inverter rated at 3200W is selected, then:

Inverter output current = 3200W ÷ 230V = 13.9A. The inverter output current is less than the maximum (16A per phase) allowed. For an optimally tilted and oriented PV array, the power at the PV inverter terminals should not be less than: 3600W × 0.9 = 3240W. Since the PV array is optimally tilted and oriented, the 3200W inverter is not acceptable (***4.1.6 Selection of grid-tied PV inverter***).

Determination of K_K (see ***Guide to the Installation of Photovoltaic Systems (2012) section 3.7.6.1***):

- PV array pitch has been calculated to be 35^0
- PV array orientation is equal to 0^0 (facing true south)
- Zone 12
- Postcode NR
- K_K = 961 kWh / kW_P from the appropriate location specific-table (***Annex D of the Guide to the Installation of Photovoltaic Systems (2012)***)
- The PV array is free of any shading (from site inspection)

Annual AC output (kWh /year) = kW × K_K × SF. So,

PV system AC output per year = 3.6 kW × 961 kWh / kW / year = 3459.6 kWh / year

Design:

A) Determine PV module area per kilowatt peak (m^2 / kW):

This has been calculated (see ***Initial PV System Energy Output Estimate*** for this example). So,

PV module area per kilowatt peak = 7.05m^2 / kW

B) Determine initial number of PV modules and PV array peak power (kW):

This has been calculated (see ***Initial PV System Energy Output Estimate*** for this example). So,

Initial number of PV modules = 22. This is the maximum number of PV modules that can fit into the available roof space; however, it is unacceptable because it gives higher peak wattage than that allowed by the tariff scheme.

The initial number of 20 PV modules will give 4000W of PV array, which is acceptable.

The 4000W of PV array is the upper limit required by the tariff scheme, because the power output at the inverter terminals would be such that the current at the inverter terminals would not exceed 16A, as shown in the ***Initial PV System Energy Output Estimate***.

The initial number of PV modules = 18 (number of PV modules selected in the initial PV system energy output estimate).

C) PV module layout:

i) To find the number of PV modules fixed in landscape format, divide length of roof (8m) by length of PV module (1.425m), and divide roof width (4m) by PV module width (0.99). So,

8m ÷ 1.425m = 5.61. That means there are five PV modules in a row.

4m ÷ 0.99m = 4.04. That means there are four rows of PV modules.

This means that the landscape format has four rows of five PV modules, i.e. 20 PV modules. This gives PV array peak power of (200 x 20) W = 4000W.

ii) To find the number of PV modules fixed in portrait format, divide length of roof (8m) by width of PV module (0.99m), and divide width of roof (4m) by length of PV module (1.425m). So,

8m ÷ 0.99m = 8.08. That means there are eight PV modules in a row.

4m ÷ 1.425m = 2.81. That means there are two rows of PV modules.

That means that the portrait format has two rows of eight PV modules, i.e. 16 PV modules. This gives PV array peak power of (200 x 16) W = 3200W

Deciding on PV module layout and number of PV modules:

The PV array roof space has taken into account the specification from the manufacturer of the PV module supporting frames, regarding distances between modules and between modules, edge of the roof, and open space for easy installation and maintenance of the PV system. In this example, 80% of the roof has been made available for the PV system. Either format of the PV module layout, landscape or portrait, can be adopted since the PV array peak wattage (W) for either is not greater than 4000W, as required by the tarrif scheme. A decision has to be made on what format to use and on the number of modules for the PV array and number of modules per PV string. The decision regarding the number of PV modules to be selected for the PV array will depend on the desired PV string voltage, the inverter MPPT voltage range, the maximum DC input voltage, and the inverter input current, as stated earlier.

Here, it is helpful to point out again that the greater the voltage for the same size of cable, the lower the voltage drop. Therefore, the PV string with the highest voltage is preferred for higher system efficiency – with reduced voltage drop and power losses.

In this example the PV array can have an equal or unequal number of PV modules in each PV string because the owner has given a go-ahead to choose either a central PV inverter or a multi-string PV inverter.

If the 20 PV modules are selected, they should be laid out in landscape format, and they could be set up into two PV strings of 10 PV modules per PV string, four PV strings of five PV modules per PV string, or one PV string of 20 PV modules. If the 16 PV modules are selected, they could be set up into two PV strings of eight PV modules per PV string, four PV strings of four PV modules per PV string, or one PV string of 16 PV modules, and they should be laid out in portrait format. If the 18 PV modules used in the initial estimate (see ***Initial PV System Energy Output Estimate***) are selected they can be set up into two PV strings of nine PV modules per PV string, three PV strings of six PV modules per string, or one PV string of 18 PV modules.

While both the 4000W and 3200W PV arrays are acceptable, the client is satisfied with the 3600W that was selected in the initial PV system energy estimate for this example. The 18 PV modules will be set up into two PV strings of nine PV modules, and the PV modules will be laid out in landscape format.

Determining PV module voltages at the extreme module temperatures:

Calculate module voltages at the extreme PV module temperatures at the site, i.e. at -10^0C and $+70^0$C:

V_{MPP} (at -10^0C) = 26.3V + (- 35 × – 0.123) V = 30.61V

V_{MPP} (at +70^0C) = 26.3V + (45 × – 0.123) V = 20.77V

V_{OC} (at -10^0C) = 32.9V + (–35 × - 0.123) V = 37.21V

The PV module maximum power point voltage will be 30.61V at the lowest site temperature (–10^0C) and will be 20.77V at the highest site temperature (70^0C). The highest PV module open-circuit voltage of 37.21V will occur at the lowest site temperature (–10^0 C).

Determination of size (kW) of inverter and inverter type selection:

Whereas the peak power of the PV module/array is obtained at standard test conditions, in practice, the PV module/array rarely operates at STC due to weather conditions, soiling, and other factors (see ***1.9 Partial Shading***). As a result, the inverter power (W) is usually lower than the PV array peak power. The selected PV inverter is rated at 3400W. The power of the PV inverter is determined as shown in the ***Initial PV System Energy Output Estimate*** for this example. Note that a PV inverter rated at 3200W cannot be recommended. Since the PV array is optimally tilted and oriented, the inverter could be overloaded (see ***4.1.6 Selection of grid-tied PV inverter***).

A PV multi-string inverter is chosen because it will greatly mitigate losses due to large PV module tolerances, and for easy PV system installation (no need for junction box, main DC cables).

The technical data sheet of this PV multi-sting inverter shows the following:

- Recommended PV power (max power of the PV array) 3660W
- DC nominal power ($P_{NOMINAL}$) 3400W
- Maximum power point tracking voltage (V_{HIGHER}) 450V
- Maximum power point tracking voltage (V_{LOWER}) 100V
- DC maximum voltage input ($V_{DC\ MAX\ INPUT}$) 500V
- DC maximum input current ($I_{DC\ MAX\ INPUT}$) 15.7A
- Euro efficiency 96.4
- Protection degree IP65
- Operating consumption 0W
- Operating temperature range -25^0C to +60^0C

Number of PV module strings:

Since the number of modules to be connected in a string is determined by the input voltage of the inverter, to find the maximum number of PV modules to be connected in each string we divide the higher limit of maximum peak power tracking voltage (V_{HIGHER}) of the inverter by the PV module maximum power point voltage, at the lowest PV module temperature at the site (V_{MPP} at -10^0C) of the PV array. So,

$$V_{HIGHER} \div V_{MPP} \text{ (at } -10^0C) = 450V \div 30.61V = 14.7 \text{ (round down)}$$

This means that the maximum number of PV modules in each PV string should not be greater than 14.

To find the minimum number of PV modules in each PV string, we divide the lower limit of maximum peak power tracking voltage (V_{LOWER}) of the inverter by the PV module maximum power point voltage, at the highest PV module temperature at the site (V_{MPP} at $+70^0C$) of the PV array. So,

$$V_{LOWER} \div V_{MPP} \text{ (at } +70^0C) = 100V \div 20.77V = 4.8 \text{ (round up)}$$

The minimum number of PV modules in each PV string is five. Therefore, for the PV array to operate within the voltage range of the inverter, the maximum number of PV modules in each PV string must not exceed 14 and the minimum number must not be less than five.

The number of PV modules in each PV string must be such that the PV string open-circuit voltage, at -10^0C, does not exceed the input voltage of the inverter. Therefore, the maximum input voltage of the inverter (**$V_{DC\ MAX\ IMPUT}$**) must be divided by the open-circuit voltage, at the lowest temperature of the PV module/array at the site (V_{OC} at -10^0C), to get the maximum number of PV modules in each PV string. So,

$$V_{DC\ MAX\ IMPUT} \div V_{OC} \text{ (at } -10^0C) = 500V \div 37.21V = 13.4 \text{ (round down)}$$

This shows that the maximum number of PV modules in a PV string should not exceed 13. The final PV module set-up specification is that the number of modules in each PV string must not exceed 13 and must not be less than five.

PV module set-up and PV array/inverter compatibility:

The chosen 18 PV modules to be set up into two PV strings of nine PV modules look acceptable because the number of PV modules in each PV string is below the maximum (13) and above the lowest (five) PV modules per PV string. We should check whether the voltages and currents of the PV array, with nine PV modules per PV string, are compatible with the input voltages and currents of the inverter.

$$V_{MPP} \text{ (at } -10^0C) = 30.61V \text{ x } 9 = 275.49V$$

The number of nine PV modules per string is acceptable since the PV string maximum power point voltage (V_{MPP}), at the lowest PV module temperature at the site, is lower than the higher limit of maximum power point tracking voltage (V_{HIGHER} = 450V) of the inverter and higher than its lower limit of maximum power point tracking voltage (V_{LOWER} = 100V).

V_{MPP} = (at +70^{0}C) = 20.77V x 9 = 186.93V

The maximum power point voltage of the nine PV modules per PV string, at the highest PV module temperature at the site, is higher than the lower limit of maximum power point tracking voltage (V_{LOWER} = 100V) of the inverter and lower than the higher limit of MPPT voltage (V_{HIGHER} = 450V). This is acceptable.

V_{OC} (at –10 ºC) = 37.2V x 9 = 334.8V

The open-circuit voltage, at the lowest site temperature (V_{OC} at –10^{0}C), of the nine PV modules per PV string is lower than the maximum DC input voltage ($V_{DC\ MAX\ INPUT}$ = 500V) of the inverter. This is acceptable.

The PV module current, at maximum power point, is equal to 7.61A (from the PV module technical data).

Since the PV array is connected to a multi-string inverter, each PV string has a current of 7.61A.

This is acceptable because the PV string current of 7.61A, at maximum power point (I_{MPP}), is below the maximum DC input current ($I_{DC\ MAX\ INPUT}$ of 15.7A) of the multi-string inverter.

The short-circuit current (I_{SC}) of the PV string is equal to the PV module I_{SC} given in the PV module technical data (at standard test conditions). So,

PV string short-circuit current (I_{SC}) = 8.21A (the PV array short-circuit current is required in the rating of string cables and DC isolators).

Final estimation of PV system energy output (KWh) per year (for this example):

PV system energy output = kWh / kW × PV array peak power × shading factor;

i) PV array peak power:

- PV array peak power is equal to 3600W.

ii) Determination of kWh / kW (K_K):

- PV array pitch has been calculated to be 35^0;
- PV array orientation is equal to 0^0 (facing true south);
- Zone 12;
- Postcode NR;
- KK = 961 kWh / kW (***Guide to the Installation of Photovoltaic Systems (2012), annex D***).

iii) Determination of shading factor) SF:

- The PV array is free of shading (from site inspection)

Therefore,

PV system AC output per year = 3.6 kW x 961 kWh / kW = 3459.6 kWh / year

The calculated PV system AC output is accompanied by a disclaimer (***Guide to the Installation of Photovoltaic Systems (2012)**, **section 3.7.8)***.

GRID-TIED PV SYSTEM LAYOUT (N.T.S)

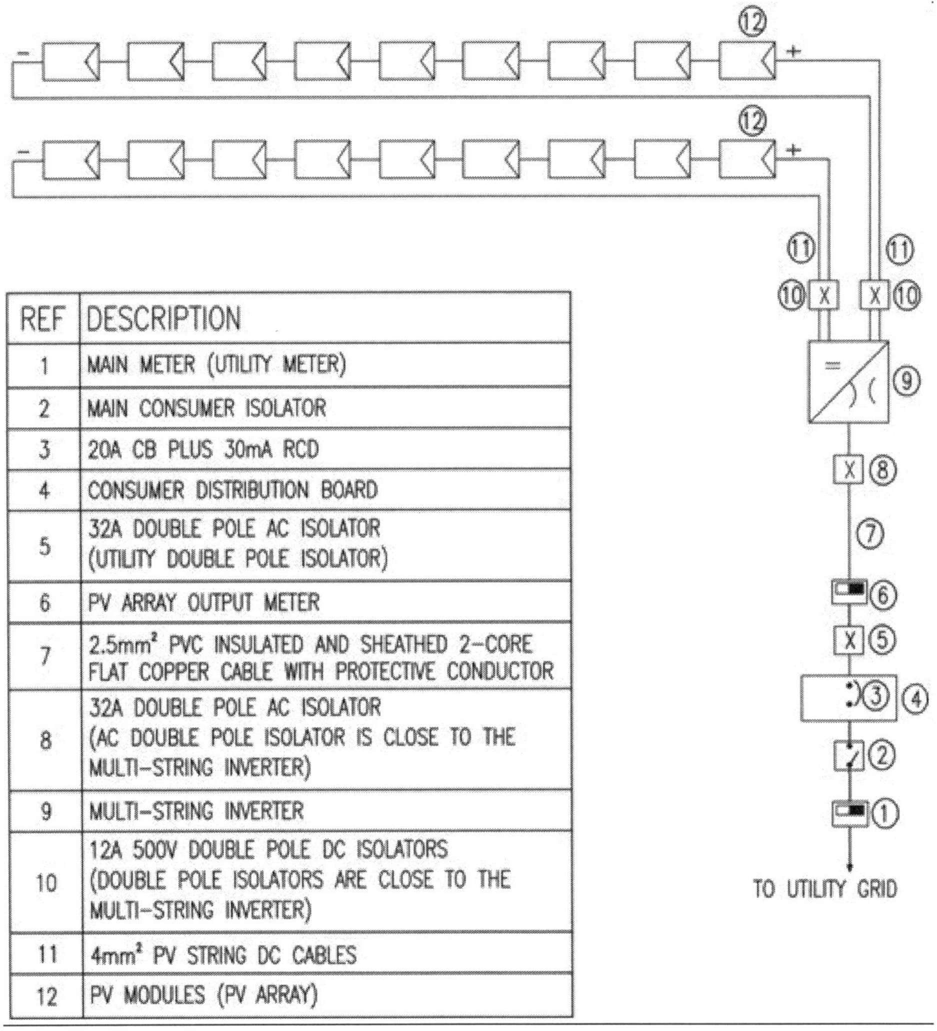

REF	DESCRIPTION
1	MAIN METER (UTILITY METER)
2	MAIN CONSUMER ISOLATOR
3	20A CB PLUS 30mA RCD
4	CONSUMER DISTRIBUTION BOARD
5	32A DOUBLE POLE AC ISOLATOR (UTILITY DOUBLE POLE ISOLATOR)
6	PV ARRAY OUTPUT METER
7	2.5mm² PVC INSULATED AND SHEATHED 2-CORE FLAT COPPER CABLE WITH PROTECTIVE CONDUCTOR
8	32A DOUBLE POLE AC ISOLATOR (AC DOUBLE POLE ISOLATOR IS CLOSE TO THE MULTI-STRING INVERTER)
9	MULTI-STRING INVERTER
10	12A 500V DOUBLE POLE DC ISOLATORS (DOUBLE POLE ISOLATORS ARE CLOSE TO THE MULTI-STRING INVERTER)
11	4mm² PV STRING DC CABLES
12	PV MODULES (PV ARRAY)

Figure 8: Grid-tied PV system layout for Example 4.

5.4 Example 7

Design a grid-tied solar electric system for an office block in Luanda, Angola. The PV array will be installed on the flat roof of the administration block. The PV system will be connected to the utility grid.

From the discussion with the client:

- Finance is limited to £65,000.
- Estimated cost per kilowatt peak is £5050 / kW.
- The PV array is to be installed on a flat roof.
- The PV installation is to be made up of two PV systems, with each PV system having a meter for measuring the PV array's electrical output.
- The owner wants to take advantage of available renewable energy incentives, and to contribute to carbon-dioxide reduction.
- The owner wants a monocrystalline PV module.
- Two PV systems are required, each with an optional inverter, i.e. either a central inverter or a multi-inverter.
- The main utility meter to the consumer distribution board will be an import meter.

From the site inspection:

- The available area for the PV array is $90m^2$ (15m long and 6m wide).
- The site is free of any shading.
- The site is at latitude 8.84^0 south and longitude 13.2894^0 east.
- PV modules will be tilted at 10^0 (though a little less than optimum tilt, this will assist PV module cleaning).
- The estimated PV module inter-row spacing is 1.13 metres.
- The PV module will be oriented toward true north
- The daily average solar radiation (kWh / m^2 / day) each year is 5.03 kWh / m^2 / day for a PV array fixed tilted at latitude year-round.
- The PV array orientation will be true north.
- Temperature at the site is between 15^0C and 40^0C.
- Temperature range of the PV module is expected to be between 15^0C and 70^0C.

Calculating array peak power:

The selected PV module is a monocrystalline silicon module, CL200 Watt, which has a technical data sheet showing the following:

Electrical characteristics (at STD):

- Maximum power (W), 200W

- Power tolerance, +3% / –3%
- Voltage at maximum power point (V_{MMP}), 37.4V
- Current at maximum power point (I_{MMP}), 5.1A
- Open-circuit voltage (V_{OC}), 45.2V
- Short-circuit current (I_{SC}), 5.4A
- Maximum system voltage 1000V
- Temperature coefficient for V_{OC} – 0.34 x $10^{-2}V/^{0}C$
- Temperature coefficient for current 0.05 x $10^{-2}A/^{0}C$
- Temperature coefficient for power, -0.455 x $10^{-2}W/^{0}C$
- Efficiency 16.4%
- Temperature rating is $-40^{0}C$ to $90^{0}C$

Mechanical characteristics:

Materials and Accessories Data (All certified by CE and TUV)

- Junction box, IP65 rated
- Type of cable connector, MC Plug Type IV
- Fuse rating 10A

Dimensions:

- Length 1580mm
- Width 808mm
- Depth 45mm

Now that we have selected the PV module we can calculate the PV array peak power as follows:

Using estimated cost per kilowatt peak:

Determine array peak power using the estimated cost of PV system per kW. Divide the available finance by the estimated cost per kilowatt peak. So,

PV array peak power = £65000.00 ÷ £5050 / kW = 12.87 kW.

Initial number of PV modules = 12870W ÷ 200W = 64.35. That means 64 PV modules. So,

PV Array peak power = 64 × 200W = 12800W

Using the available area for the PV array:

Divide the available area by the PV module area per kilowatt peak to obtain the PV array peak power. PV module area per kilowatt peak = 1.580m x 0.808m x 5 / kW = 6.38m^2 / kW. So,

PV array peak power = 90m^2 ÷ 6.38m^2 / kW ÷ space factor = 14.11 kW ÷ 1.13 = 12.49 kW

The 12.49 kW is acceptable because it is less than the 12.87 kW determined through estimated cost per kilowatt peak.

Module peak power is 200W (see PV module technical data). The initial number of PV modules is calculated as follows:

12490W ÷ 200W = 62.45. This means that the initial number of PV modules for the PV array is equal to 62 PV modules. So,

PV array peak power (initial W_P) = 200 W × 62 = 12400 W

PV module layout:

Landscape format:

To determine number of PV modules per row divide the length of site area (15m) by length of PV module (1.58m). So,

15m ÷ 1.58m = 9.49. That means there are 9 PV modules in a row.

To determine the number of rows of PV modules divide the width of the site area (6m) by the width of the PV module (0.808m) and divide by the space factor (1.13). So,

6m ÷ 0.808m ÷ 1.13 = 6.7. This means that there are six rows of PV modules.

Therefore, the PV array will have six rows of nine PV modules per row (6 x 9 = 54 PV modules) that should be laid out in landscape format. This gives a PV array peak power of 10.8 kW.

Portrait format:

To determine the number of PV modules in a row divide the length of the site area (15m) by the width of the PV module (0.808m). So,

15m ÷ 0.808m = 18.56. This means that there are 18 PV modules in a row.

To determine number of rows of PV modules divide the width of the site area (6m) by the length of the PV module (1.58m) and divide by the spacing (1.13). So,

6m ÷ 1.58m ÷ 1.13 = 3.36. This means that there are three rows of PV modules.

Therefore, the PV array will have three rows of 18 PV modules per row (3 x 18 = 54 PV modules) that should be laid out in portrait format. This gives PV array peak power of 10.8 kW.

The 54 PV modules laid out in landscape format and the 54 PV modules laid out in portrait format both have peak power that is less than the peak power obtained through the estimated cost. Therefore, the 54 modules laid out in either format are acceptable. The 54 modules are divided into two PV arrays, each for a separate PV system. The PV modules must be set up into PV strings with an equal number of modules per PV string. They could be set up into three PV strings of 18 modules per string, or six PV strings of nine PV modules per string. In the example, the PV modules will be laid out in landscape format.

***Calculating PV module voltage at the extreme module temperatures*:**

$$V_{OC} \text{ (at } 15^0\text{C)} = 45.2\text{V} + (-\ 10 \times (-\ 0.34 \div 100))\ \text{V} = 45.23\text{V}$$

$$V_{MPP} \text{ (at } 15^0\text{C)} = 37.4\text{V} + (-\ 10 \times (-\ 0.34 \div 100)\ \text{V} = 37.43\text{V}$$

$$V_{MPP} \text{ (at } 70^0\text{C)} = 37.4\text{V} + (45 \times (-\ 0.34 \div 100)\ \text{V} = 37.25\text{V}$$

The maximum power point voltage range is between 37.25V and 37.43V, and the highest open-circuit voltage is 45.25V.

Inverter selection:

A central inverter is selected. Since two PV systems, each connected to a central inverter, are required, the peak power of each PV array will be equal to 5.4 kW. Inverter nominal power should be slightly less than the PV array peak power, and, in this example, the inverter input power should not be less than 90% of the PV array's peak power. The selected inverter has a nominal DC power of 5.3 kW. The technical specification data sheet of this inverter shows the following:

- Recommended PV power (at PV module STC) 6 kW
- DC nominal power ($P_{DC\ NOMINAL}$) 5.3 kW
- Maximum peak power tracking voltage (V_{HIGHER}) 500V
- Minimum peak power tracking voltage (V_{LOWER}) 100V
- Maximum DC input voltage ($V_{DC\ MAX\ INPUT}$) 550V

- Maximum DC input current ($V_{DC\ MAX\ INPUT}$) 25A
- Nominal output AC current 20A
- Maximum output AC current 23.8A

Number of PV modules per string:

The number of PV modules in a PV string is determined by the input voltage of the inverter. Therefore, to find the maximum number of PV modules to be connected in each PV string, divide the higher limit of the maximum peak power tracking voltage (V_{HIGHER}) of the inverter by the PV array maximum power point voltage (V_{MPP}), at the lowest PV module temperature at the site (V_{MPP} at 15^0 C). So,

$$V_{HIGHER} \div V_{MPP} \text{ (at } 15^0\text{C)} = 500\text{V} \div 37.43 = 13.36 \text{ (round down)}$$

This means that the PV modules in each PV string should not exceed 13 PV modules.

To find the minimum number of PV modules in each PV string, divide the lower limit of the maximum peak power tracking voltage (V_{LOWER}) of the inverter by the PV array's maximum power point voltage at the highest PV module temperature at the site (V_{MPP} at $+70^0$C). So,

$$V_{LOWER} \div V_{MPP} \text{ (at } +70^0\text{C)} = 100\text{V} \div 37.25\text{V} = 2.68 \text{ (round up)}$$

This means that the PV modules in each string must not be less than three.

Therefore, for the array to be within the range of the inverter, that is, the range in which the inverter will be able to track the PV array's maximum power point, the maximum number of PV modules in each PV string should not be more than 13 and the minimum should not be less than three.

As already stated, the maximum voltage of the PV array will take place at the inverter input terminals during open-circuit voltage (V_{OC}) when the temperature at the site is lowest, at the instant the PV array becomes exposed to sunlight. The number of PV modules in each PV string must be such that the PV string voltage does not exceed the input voltage of the inverter. Therefore, the maximum input voltage ($V_{DC\ MAX\ INPUT}$) of the inverter must be divided by PV array open-circuit voltage at the lowest site temperature (V_{OC} at 15^0C). So,

$$V_{DC\ MAX} \div V_{OC} \text{ (at } 15^0\text{C)} = 550\text{V} \div 45.25\text{V} = 12.16 \text{ (round down)}$$

This means that the highest number of PV modules that can be connected in each PV string should not exceed 12 PV modules so that the output voltage of the PV array will never exceed the input voltage of the inverter.

The PV module set-up and PV array/inverter compatibility:

As stated earlier, the 54 PV modules laid out in landscape format and the 54 PV modules laid out in portrait format are both acceptable because their respective peak wattage is less than the peak power (12.87 kW) obtained through estimated cost per kilowatt peak. The PV modules must be set up in such a way that the PV modules per PV string give maximum power point voltage (V_{MPP}), at extreme PV module temperatures, which lies within the inverter MPPT voltage range, and give the PV string open-circuit voltage at the lowest site temperature, which is below the maximum DC input voltage of the inverter. Therefore, the selected number of PV modules should be set up in such a way that the PV modules in each PV string are not less than three and not more than 12. Additionally, since a PV central inverter has been selected, the PV strings should have the same number of modules.

The 54 PV modules can be set up into six PV strings of nine PV modules per PV string. The nine PV modules per PV string are below the maximum (12 modules) and above the minimum (three modules) required for each PV string. Therefore, the initial number of 54 PV modules made up of six PV strings, with nine PV modules per PV string, is selected, and the PV modules must be laid out in landscape format. The 54 PV modules are subdivided into two PV arrays of three PV strings with nine PV modules per string, and each PV array is connected to a separate central inverter. This means that the installation will be made up of two PV systems, both connected to the consumer distribution board.

Now check whether the voltages and currents of the PV array are compatible with the input voltages and currents of the relevant inverter.

$$V_{MPP} \text{ (at } 15^0\text{C)} = 37.43\text{V x } 9 = 336.87\text{V}$$

Since the PV string's maximum power point voltage (V_{MPP} = 336.87) at the lowest PV module temperature at the site is lower than the higher limit of the maximum peak power tracking voltage (V_{HIGHER} _500V) of the inverter and higher than the lower limit of the maximum power point tracking voltage (V_{LOWER} _ 100V) of the inverter, this is acceptable.

$$V_{MPP} \text{ (at } +70^0\text{C)} = 37.25\text{V x } 9 = 335.25\text{V}$$

Since the PV string maximum power point voltage (V_{MPP}), at the highest site temperature, is higher than the lower limit of the maximum peak power tracking voltage (V_{LOWER} _ 100V) and lower than the higher limit (V_{HIGHER} _ 500V) of the inverter, this is acceptable.

$$V_{OC} \text{ (at } 15^\circ\text{C)} = 45.23\text{V x } 9 = 407.07\text{V}$$

Since the PV string open-circuit voltage (V_{OC}) is lower than the maximum DC input voltage ($V_{DC\ MAX\ INPUT}$ _ 550V) of the inverter, this is acceptable.

The PV module current at maximum power point (I_{MPP}) is equal to 5.1A (see PV module technical data). Since the 54 PV modules are divided equally between two PV systems, each connected to a central inverter that means each PV system has three PV strings connected in parallel in the junction box. So,

$$I_{MPP} = 5.1A \times 3 = 15.3A$$

This is acceptable because the current at maximum power point (I_{MPP}) is below the 25A maximum DC input current ($I_{DC\ INPUT}$) of the inverter.

For other PV system components on the DC side see ***4.2 Other DC Components for Grid-Tied PV System***.

REF	DESCRIPTION
1	MAIN METER (IMPORT - EXPORT METER)
2	MAIN CONSUMER ISOLATOR
3	CONSUMER DISTRIBUTION BOARD
4	MINIATURE CIRCUIT BREAKERS (MCBs)
5	PV ARRAY OUTPUT METERS
6	A.C. CABLES (A.C. CABLES TO CONSUMER DISTRIBUTION BOARD)
7	A.C. DOUBLE POLE ISOLATOR (THE D.C. DOUBLE POLE ISOLATOR IS CLOSE TO THE INVERTER)
7A	A.C. DOUBLE POLE ISOLATOR (UTILITY ISOLATOR)
8	GRID-TIED CENTRAL INVERTER
9	D.C. DOUBLE POLE ISOLATOR (D.C. ISOLATOR IS PART OF THE INVERTER)
10	MAIN D.C. CABLES
11	D.C. JUNCTION BOX (D.C. JUNCTION BOX IN WHICH THE PV STRINGS ARE CONNECTED IN PARALLEL)
11A	PV STRING FUSES
12A	PV STRING CABLES (POSITIVE)
12B	PV STRING CABLES (NEGATIVE)
13	PV MODULES (PV ARRAY)

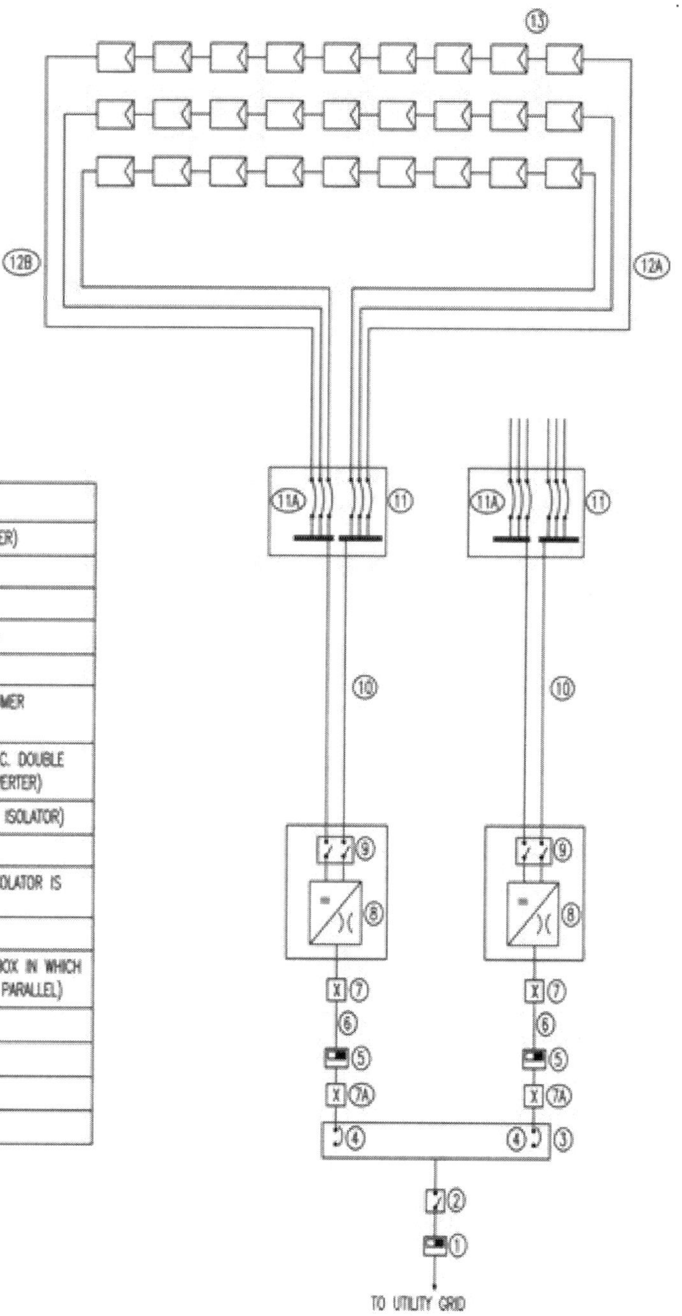

Figure 9: Grid-tied PV system layout for Example 5

6. Standalone PV Systems

6.1 Overview

As the term implies, a standalone PV system is a PV system that is not connected to the utility grid. It is also known as an *autonomous* or *island* PV system or an *off-grid* PV system. Standalone PV systems are usually installed in remote areas (where the utility grid is not available or where the connection to the utility grid is much more expensive than the installation of a standalone PV system).

Most standalone PV systems have a storage facility, i.e. the PV module/array is connected to the load via a storage facility such as a battery/battery bank. Standalone PV systems range from very small PV systems of about 0.1 watt, such as the PV system for pocket calculators, to about 10 kilowatts for larger PV systems such as PV systems for DC and AC pumps, and to mega-watts for hybrid PV systems.

The standalone PV system can be subdivided into hybrid and non-hybrid systems. Non-hybrid PV systems have a PV generator but have no additional energy source. The hybrid PV systems have, in addition to the PV array (PV generator), another source or other sources of supply of electrical energy, which may be a wind, diesel, gas, or bio-fuel generator, connected in the system. Our discussion here will exclude very small PV systems, such as PV systems for wristwatches and lanterns. Here, we will be concerned particularly with PV systems of between 50 and 2500 watts. However, some discussion will be made on hybrid PV systems.

Cost-wise, it would not be a good idea to install a standalone PV system in locations where the utility grid is available most of the time or where it is likely to be available sooner rather than later. Standalone PV systems can be very necessary (provided the location at which the PV system is installed has plentiful sunshine) in areas where there is no utility grid, such as remote areas in the developing world, or where there are frequent long electric power outages from the grid.

A standalone PV system may be used to provide electricity to the following, amongst others:

- Pumps for pumping water for drinking or for livestock, or for small vegetable gardens or irrigation (in remote areas);
- Schools (in remote areas, mostly in the developing world);
- Clinics (in remote areas, mostly in the developing world);
- Telecommunication systems, security systems, medical refrigeration, bus stop shelter lighting;
- Holiday dwellings, shelters (in remote areas, mostly in the developing world);
- Rural homes (mostly in the developing world).

One example will be given concerning the design of a standalone PV system for a rural home. A brief description of different layouts for standalone PV systems and PV hybrid systems will also be made. PV modules, inverters, and batteries for standalone PV systems and battery charge controllers will be discussed, and brief reference will also be made to other components of the standalone PV system, e.g. DC-DC converter, bypass diodes, blocking diodes, etc.

As stated earlier, the majority of PV arrays for standalone PV systems will be connected via a battery or battery bank to the load. This is because the sun's energy falling on the PV array is not constant (see ***2.1 Capturing of the Sun's Energy***). This means that, in order to supply a constant load and to meet the energy shortfall from the PV array, on cloudy days or during the night, a storage facility is needed. When there is sunlight there will be flow of electrical energy from the PV array to the battery and from the battery to the load. If for any reason there is less energy withdrawn by the load, there is a possibility of the battery being over-charged by the PV array. This means that there should be a way of preventing battery over-charge. On the other hand, the energy stored in the battery can be depleted over time to below acceptable level (to which the battery is designed), e.g. when the PV array cannot replenish the electrical energy being withdrawn by the load. The failure to replenish the energy withdrawn by the load may be due to incorrect PV array/battery set-up or a fault in the PV system. Thus, the battery in a standalone PV system should always be protected against over-charge or over-discharge, and against overload and short-circuit (see ***6.5.2 Battery for standalone PV system***). The protection against over-charge and over-discharge of the battery is provided by connecting a battery charge controller (also known as a regulator) between the PV array and the battery (see ***6.5.3 Battery charge controller***). The protection against overload and current short-circuit is provided by a fused switch or circuit breaker between the charge controller and the battery, and between the battery and the DC load and or the inverter. Means of isolating the PV array from the rest of the system should be provided by a fused disconnection switch, and all the circuits to loads must be protected by a fuse or circuit breaker. Usually, a battery charge controller has a mechanism for preventing the electrical current from flowing from the battery to the PV array during the night, and where the controller cannot provide protection against reverse current, a blocking diode should be provided. For smaller standalone PV systems, only one blocking diode should be provided on one main outgoing PV string cable (see **Figure 11**). In larger PV systems with several PV strings, blocking diodes should be provided for each PV string (PV string fuses may be used in place of blocking diodes) to prevent the flow of electrical current from healthy PV strings into PV strings with shaded or failed PV modules (see **Figure 6** and **Figure 10**). In addition, systems with several PV modules in each PV string, bypass diodes may be provided if they are required. Refer also to ***1.10 PV Modules Set-Up and Array Output***.

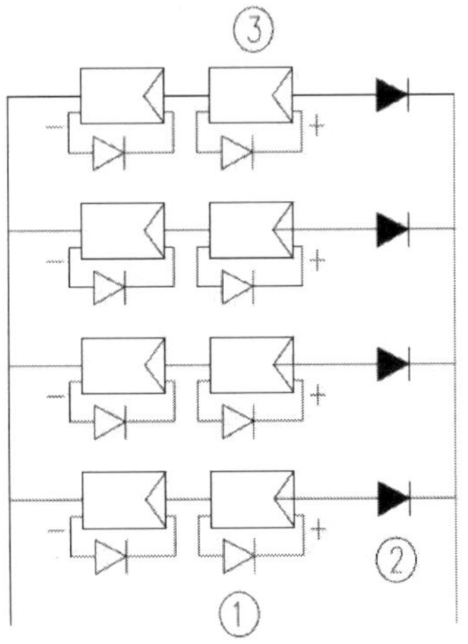

REF	DESCRIPTION
1	BYPASS DIODE
2	BLOCKING DIODE
3	PV MODULE

Figure 10: Bypass diodes and blocking diodes in larger standalone PV systems

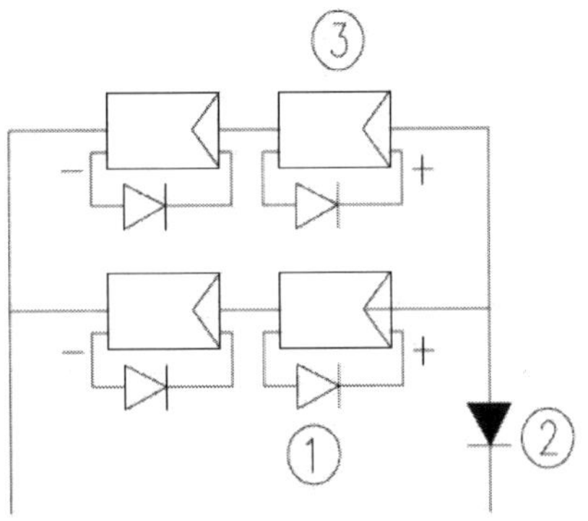

REF	DESCRIPTION
1	BYPASS DIODE
2	BLOCKING DIODE
3	PV MODULE

Figure 11: Bypass diodes and blocking diodes in small standalone PV systems

6.2. Various Standalone PV System Configurations

6.2.1 Non-hybrid system

PV array connected direct to DC load (non-hybrid system):

A standalone PV array can be connected direct to the DC load. However, the connection should be via a fused isolation switch for isolating the PV array from the rest of the system. In addition, short-circuit protection, overload protection, and control switches should be provided as required. Depending on the size of the load, where the load is a motor, a manual or automatic power control to the load may be required. The PV array voltage should match the DC load voltage. The PV system will operate only during the day when there is sunlight. Where the load is a motor, the motor speed will increase with the increase of sunlight and vice versa. This PV system arrangement is suitable for small DC electric motors such as ventilation fans, small circulation pumps for thermal water heating systems, water pumps without water storage tank for homes, and is also used for larger pumps with a water storage tank for homes, and pumps for irrigation or livestock watering (see **Figure 12** below). For larger water pumps for homes or water pumps for irrigation or livestock watering, a controller should be provided. This controller is an electronic device that controls the operation of the load (pump motor) and usually the controller has a disconnection switch for isolating the PV array from the rest of the PV system. The controller should switch off power to the pump when the power is not required by the load, and vice versa. Where the PV system is for pumping water the controller will be required to: limit the power to the pump automatically, to prevent the pump motor from running beyond its rated speed when there is over-supply of electrical current from the PV array; and switch off the pump automatically when water in the water source reaches minimum level (low water dry protection), to prevent damage to the pump motor due to over-heating. Other switches may be required, such as a float switch in the storage tank or water pressure switch in the system, either of which should operate automatically. The operation of each relevant switch will switch off the pump when the water in the storage tank is full and when a designated water pressure in the system is reached, respectively, and vice versa. The manual control switch of the pump motor should override the automatic control switches.

Figure 12: PV array connected direct to DC load

PV array connected to DC load via a DC-DC converter (non-hybrid system):

A standalone PV array may be connected via a DC-DC converter to a DC load. The DC-DC converter converts the standalone PV array output voltage to DC load voltage. Isolation, short-circuit protection, and overload protection switches should be provided, as explained earlier, and where the load is a motor, e.g. for domestic water pump with a storage tank, or water pump for irrigation, a controller should be installed between the PV array and the DC-DC converter to provide automatic control of the operation of the load. This system will operate during the day only when there is sunlight. Where the load is a DC motor, the speed of the motor will increase with the increase of sunlight, and vice versa. Note that the DC-DC converter may have a maximum power point tracker (MPPT), which is essential in tracking the maximum power point (MPP) of the PV array. This arrangement (see **Figure 13**) is suitable for larger DC water pumps, with or without water storage tank, for drinking water, livestock watering, or irrigation in remote areas.

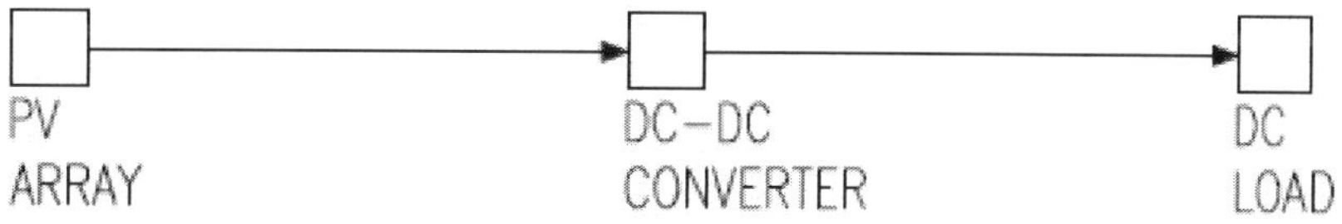

Figure 13: PV array connected to DC load via a DC-DC converter

PV array connected via an inverter to an AC load (non-hybrid system):

A standalone PV array may be connected via an inverter to AC load. Isolation, short-circuit protection, overload protection, and control switches should be provided as required. Where the PV array is connected via an inverter the output voltage and current of the PV array must be compatible with the input voltage and input current of the inverter. The inverter should be capable of tracking the maximum power point of the PV array, as already explained in ***Grid-Tied PV Systems*** (see **5.1.4 Design stage – PV module/array and inverter compatibility**). This PV system will operate during the day only when there is sunlight, since it has no storage facility. This arrangement is suitable for driving deep-well and larger AC water pumps, with or without a water storage tank, for drinking water, livestock watering, or irrigation in remote areas. Again, note that the motor speed of the pump will increase with the increase of sunlight. The inverter voltage should be suitable for the load voltage, usually 230V. The inverter DC input voltage will determine the output voltage of the PV array, and the inverter must be capable of protecting itself against variations of PV array voltage exceeding the limits of the inverter voltage range. Usually, a controller is provided to automatically limit the PV array output power to pre-set maximum and minimum values, and to provide automatic control of the load operation as explained earlier. The arrangement of the PV system will be as shown in **Figure 14** below.

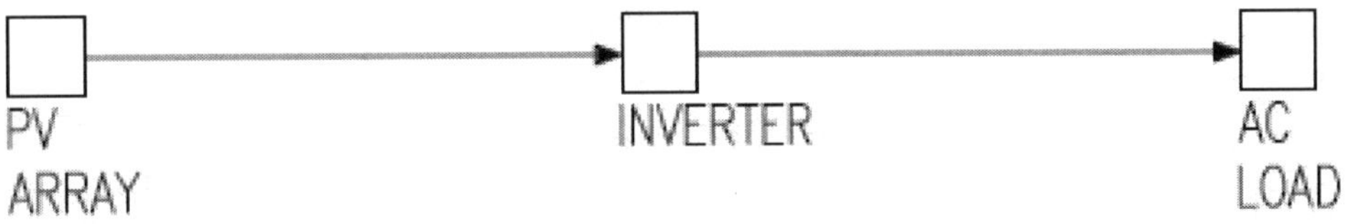

Figure 14: PV system, without a battery, supplying power to AC load

PV array connected to DC load via a battery (non-hybrid system):

A standalone PV system can have DC load connected to the battery. The battery voltage will determine the PV array voltage and must be suitable for or match the load voltage. This PV system will operate continuously, including during the night and on cloudy days (see ***6.5.2.1 Battery capacity*** and ***6.5.2.2 Autonomous days and system availability***). This is suitable for supplying small DC loads such as DC lights, radio, *small* refrigerator, or DC water pumps (for drinking water) where there is no water storage tank. When electrical power from the PV array exceeds electrical power requirements of the pump, the excess current is used for charging the battery, and when the pump requirement exceeds the solar input power to the pump, the battery powers the pump. Overload and short-circuit protection and means of isolation should be provided between the battery and the charge controller and between the battery and DC loads. A manual control switch will be required for a DC water pump. This type of PV system is suitable for homes in remote areas (especially in the developing world). The arrangement of the PV system will be as shown in **Figure 15** below.

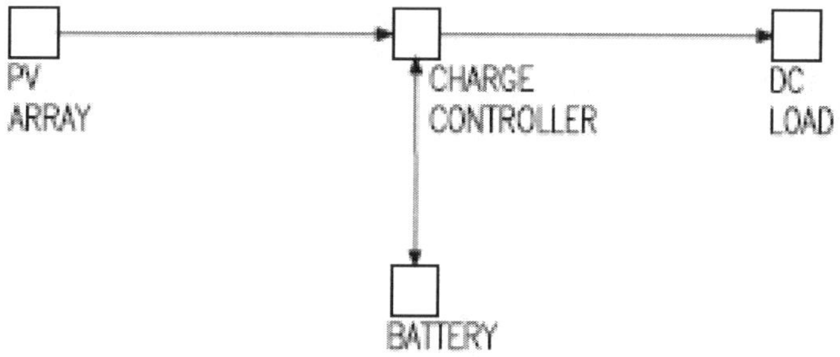

Figure 15: PV system, with a battery, supplying DC loads

PV system supplying both DC and AC loads (non-hybrid system):

A standalone PV system can have a battery and an inverter, as shown in **Figure 16**. This PV system can supply both DC and AC loads. The size (wattage) of the inverter will determine the PV system voltage. The PV system will operate continuously, even at night and during cloudy days (see ***6.5.2.2 Battery capacity and autonomous days and system availability***). The electrical power from the PV array to the battery is used to power the connected load, and excess electrical power is used to charge the batteries. Means of isolating the PV array should be installed between the PV array and the charge controller (fused connection switch, circuit breaker, or isolation switch). Additionally, overload and short-circuit protection, and means of isolation, should be provided between the battery and the charge controller, between the battery and DC loads, and between the inverter and AC loads. A PV string fuse or blocking diode may be required for each PV string depending on the number of parallel PV strings for the PV array (see **Fig 10** and **Fig. 11**). This arrangement is suitable for supplying electrical power to small homes, clinics, and schools in remote areas.

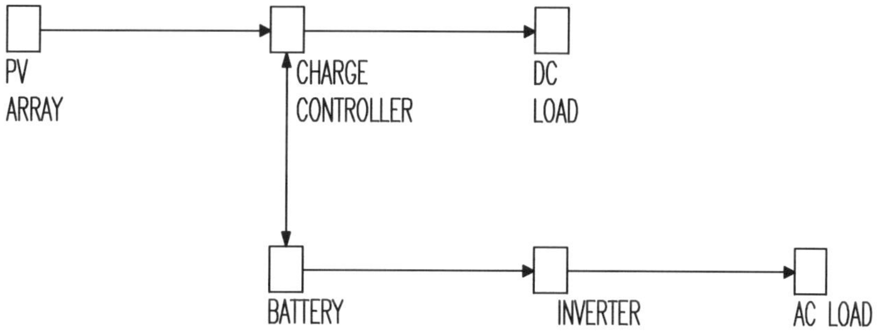

Figure 16: PV system with a battery supplying DC and AC loads

6.1.2 PV hybrid system

As stated earlier, a standalone PV system with battery, inverter, and additional power source/s is known as a hybrid PV system. Hybrid system voltage is usually 24VDC and above. The battery bank voltage should match the inverter maximum input DC voltage. The power rating of the inverter will determine the hybrid system's DC voltage, which is equal to the battery/battery bank voltage. A DC load that does not match the battery voltage (system VDC) may be connected through a DC-DC converter.

It is suggested that a hybrid PV system will be required where the load is greater than 2.5 kW or where the ratio of the PV array power to load power is greater than 0.3. This is done in order to improve the efficiency of the PV system and reduce the PV system cost. For example, a PV generator and wind generator connected in parallel will produce more energy than when working individually. The PV generator will produce more energy in summer, when there is more sunlight, than in winter, while the wind generator will produce more energy in winter, when there is more wind, than in summer. This arrangement (hybrid system) is for supplying larger loads, e.g. larger rural homes, clinics, hospitals, and schools in remote areas (especially remote areas in the developing world).

PV generator and wind generator configuration (hybrid system):

Figure 17 shows a hybrid system in which: the two generators (PV array and wind generator) are each connected to a charge controller, and the two charge controllers are connected in parallel to the battery bank; the DC loads are connected to the battery bank; and the inverter is connected to the battery bank and to the AC load. Isolation, short-circuit protection, overload protection, and control switches should be provided as required.

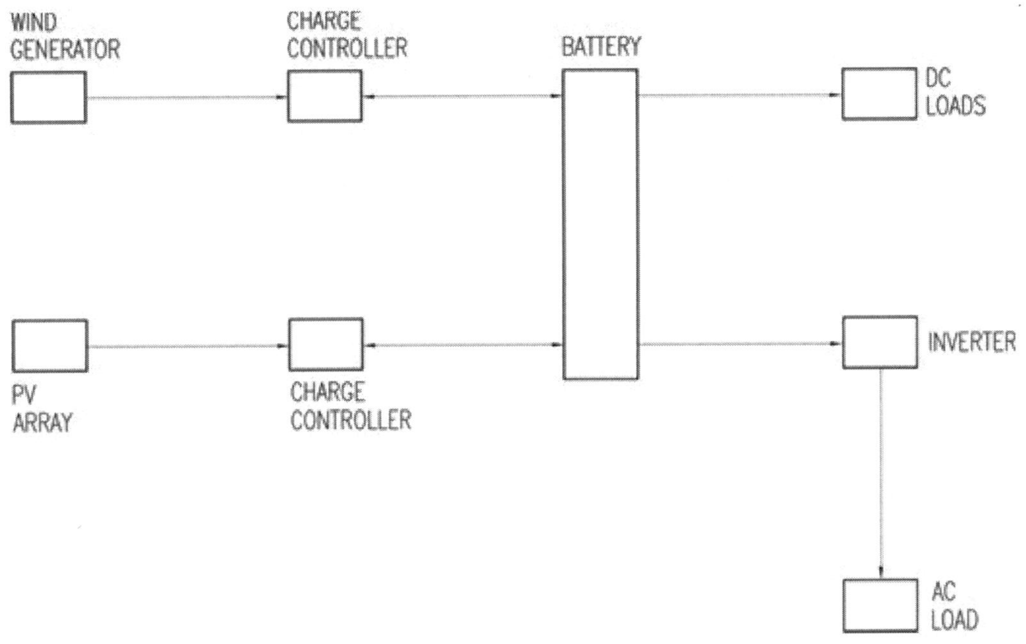

Figure 17: PV hybrid system with PV generator and wind generator

PV generator and diesel engine generator configuration (hybrid system):

Figs. 18, **19**, and **20** show three types of hybrid systems with two sources of power supply, namely PV array and diesel generator (PV array and diesel generator hybrid configuration by C. V. Nayar et al. (1993)).

Fig. 18 shows a switched hybrid system in which: the PV array is connected to the battery bank via a charge controller; the inverter is connected to the battery bank and connected via a change-over switch to the AC load; the diesel generator is connected to a battery bank bus (DC bus) via a battery charger. Consequently, when power from the diesel generator to the AC load is fed via the inverter, the current from the diesel generator is first converted by the battery charger from AC to DC current and is later converted by the inverter to AC (series connection). The diesel generator should switch on when the PV array cannot fully charge the battery bank due to a higher load demand than the power the PV array can supply (i.e. when there is no excess power from the PV array to charge the battery bank). Therefore, the inverter should be rated for the combined power of the PV array and diesel generator. The inverter rated power will determine the battery bank voltage. The diesel generator can be switched direct to AC load when AC load demand is too high to be met by the PV power supply (there is no possibility of parallel operation of inverter and diesel generator). Thus, the diesel generator should also be designed for maximum power demand. Isolation, short-circuit protection, overload protection, and switches should be provided as required. The switching on/off of the diesel generator can be done manually or automatically.

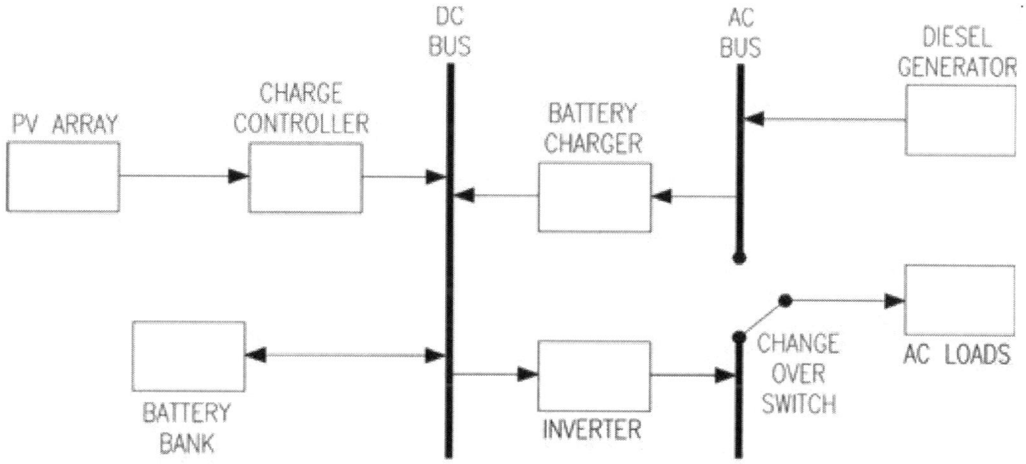

Figure 18: Switched hybrid system

Fig. 19 shows a series hybrid system in which: the PV array is connected to the battery bank via the charge controller, and the battery bank is connected to DC load (DC bus) and connected via the inverter to AC load; the diesel generator, via a battery charger, is connected to the battery bank. This means that the current from the diesel generator is first converted by the battery charger from AC to DC and later converted by the inverter to AC. The PV array, via a charge controller, charges the battery and supplies primary duty power. The diesel generator should switch on when the PV array cannot fully charge the battery bank due to a higher load demand than the power the PV array can supply. This means that the diesel generator, via a battery charger, charges the battery and provides an equalisation charge when there is not adequate sunlight, and provides power to specific loads that cannot be powered by the PV system (alone). The inverter should be rated for the combined power of the PV array and diesel generator. In this type of hybrid system the diesel generator should be sized to operate optimally. Isolation, short-circuit protection, overload protection, and switches should be provided as required. The operation of the diesel generator can be controlled manually or automatically.

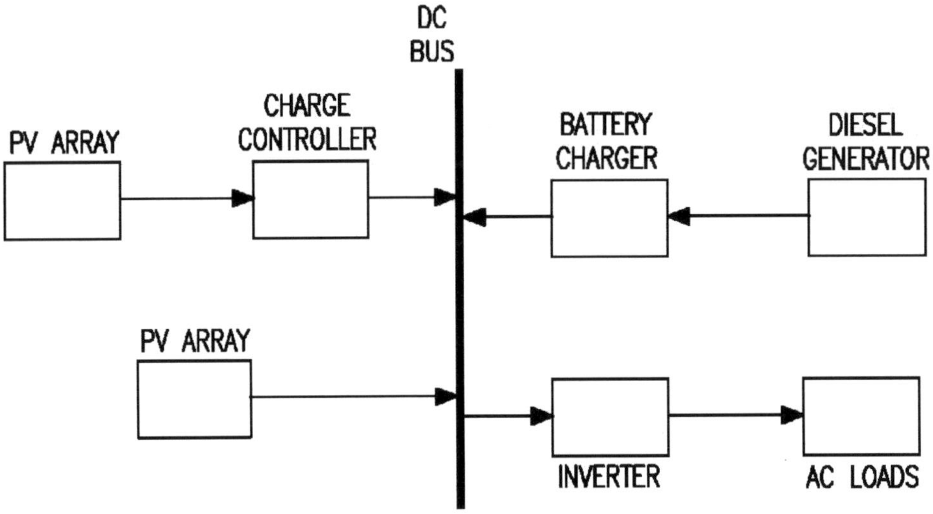

Figure 19: Series hybrid system

Fig. 20 shows a parallel hybrid system in which: the PV array is connected to the battery bank via charge controller, and the battery bank is connected to the DC load (DC bus) and via the bi-directional inverter to the AC load; the diesel generator is connected to the battery bank via a bi-directional inverter and connected directly to the AC load. The two power sources can operate individually and in parallel. When the two electrical sources supplying power to the AC load are operating in parallel, synchronisation of the inverter and the diesel generator will be required. The combined power of the inverter and the diesel generator should be capable of meeting the maximum load demand. When the diesel generator is operating, the bi-directional inverter will charge the battery bank if the battery bank is not fully charged. Isolation, short-circuit protection, overload protection, and switches should be provided as required in the system. Automatic control of the operation of the diesel generator will be required.

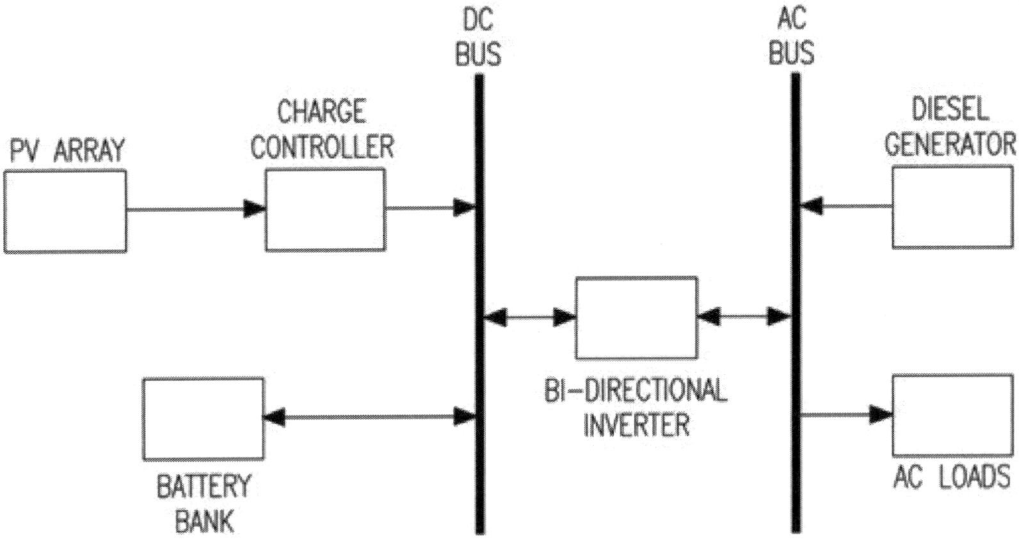

Figure 20: Parallel hybrid system

6.3 Solar Resource for Standalone PV System

Required irradiation (sun-hours) should be given for a specific location as average irradiation **per month or per day each month**, at a tilt angle for the location site. Where irradiation for the location is given as radiation on a horizontal surface, the PV module optimum tilt angle for the design month at the location site should be calculated and the irradiation adjusted for the tilt angle (see also ***2.5 Solar Resource***). The design month for standalone PV systems is usually the month with minimum irradiation per day (sun-hours per day) and maximum electrical energy demand per day.

Where the standalone PV system is required to meet electricity for the summer months, the summer month with least sun-hours should be selected as the design month. If the difference in the monthly sun-hours values of the summer months is small, the average sun-hours for the summer months can be calculated and used instead. For example, this is applicable where electricity is required for irrigation or stock watering. If the PV system is required to meet all the electricity required each day throughout the whole year, the month (winter month) with the lowest sun-hours per day must be selected as the design month. Irradiation obtained from local sources is the best. For more information refer to ***2.5 Solar Resource***.

6.4 Electrical Energy Estimate for Standalone PV System

6.4.1 Estimating PV system energy output for standalone PV system

The estimated electrical energy output for a standalone PV system is the total watt-hour, for both DC and AC load, divided by the product of battery and wiring efficiencies. For example, the watt-hours required for a 300W (AC) computer, with efficiency of 0.75, duty cycle of two hours per day, and used seven days a week will be: 300W ÷ 0.75 × 2 h / day × (7÷7) = 800W-h / day. To obtain the energy produced per day by the PV array, the total watts per day, as calculated above (namely, the 800W-h / day), should be divided with the product of the battery and cable efficiency factors. This is in line with the design method used in ***6.7.2 Example 8***. In another method to obtain the energy produced per day, the total watt-hours per day, calculated as above, would be divided by the product of battery, inverter, and wiring efficiency factors.

6. 5 Main Components for Standalone PV System

6.5.1 PV module for standalone PV system

The PV cell/module technology, and likewise the electrical characteristics of PV cell/module, for standalone PV systems is the same as that of PV cell/module for grid-tied PV systems. However, in standalone PV systems, PV modules do not usually operate at maximum power point, except for larger PV systems. In other words, in most standalone PV systems usually there is no maximum peak power tracking of the PV module maximum power point. In fact, the PV modules for a standalone PV system usually work at reduced voltage, which is sufficient to charge the battery. For example, a PV module/array connected to a 12 VDC lead acid battery should have at least output voltage of 14.4 VDC to enable the PV module/array to charge the battery. This means that the charging voltage should not be less than 120% of the nominal battery voltage. To charge a battery with a voltage rating of 24 VDC the same PV array will need be set up to produce twice that voltage, i.e. 14.4 VDC × 2. The voltage of a crystalline PV cell is 0.5 VDC to 0.6 VDC (see ***1.1 Photovoltaic Cell/Module Technology***), which means a PV module made up of 36 cells will produce about 20 VDC. After accounting for PV module mismatch losses, diode and wiring losses, charge controller losses, and voltage losses due to PV module temperature at the site, the module output voltage of 20 VDC will be reduced to about 14.4 VDC, which is the required voltage for charging a 12V battery. So, to charge a 24Volt battery bank (two 12V batteries, four 6V batteries, or twelve 2V batteries connected in series), double the number of cells (36 cells × 2), connected in series, will be required. The PV module maximum output voltage (output voltage for PV modules for standalone PV systems) is usually given in the PV module technical data sheet. If the PV module output voltage is not given in the PV module technical data sheet it should be obtained from the manufacturer. During the design of standalone PV system care should be taken to ensure that the required battery charging voltage is maintained.

In larger PV systems that have no electrical energy storage facility (battery), e.g. for supplying AC power to AC pumps, the inverter will track the maximum power point (MPP) of the PV module/array. In larger PV systems with a battery, the charge controller may provide tracking of the MPP of the PV array. For a PV system with battery/battery bank, the output voltage of the PV module/array and that of the battery/battery bank, which is usually 12 VDC, 24 VDC, or 48 VDC, is determined by the inverter rated power.

PV modules for standalone PV systems usually have junction boxes with connectors, instead of flying leads with plugs and sockets as for PV modules for grid-tied PV systems, for connecting PV module to PV module, and PV module to load. As already stated, the usual system voltage for standalone PV systems is 12V, 24V, and for larger PV systems 48V. This means that, in standalone PV systems, PV arrays usually have fewer PV modules in series and more PV modules in parallel.

Number of PV modules in series = battery (PV system) voltage ÷ PV module voltage at highest PV module temperature at the site.

Number of PV modules in parallel = PV array current (I_{MPP}) ÷ PV module (or PV string) current (I_{MPP}). Note that the current at maximum power point (I_{MPP}) is at standard test conditions.

The PV module must have a technical specification data sheet showing, among other things, the following at standard test conditions:

- V_{MPP} (voltage at maximum power point) at standard test conditions
- I_{MPP} (current at maximum power point) at standard test conditions

- $I_{SC\ STC}$
- $V_{OC\ STC}$
- Efficiency
- Maximum output voltage (for charging the battery)
- Dimensions
- Weight

6.5.1.1 Selecting a PV module for a standalone PV system

When selecting a PV module for a standalone PV system the following should be considered:

- PV module efficiency: this will take into account energy requirements versus available space and cost of the PV module (see ***1.6 PV Module Peak Power and Efficiency***).
- PV module output voltage: this should be sufficient to charge the battery (see ***6.5.1 PV Module for Stand-alone PV System***).
- PV module current, i.e. I_{MPP} and I_{SC}.
- Weight and dimensions of the PV module.
- Means for connection and durability of connectors.

Most of the information given in ***4.1.2 Selection of PV module for grid-tied PV system*** should also be taken into account.

6.5.1.2 Tilt and orientation angles in standalone PV system

PV modules in standalone PV systems should be optimally tilted and oriented (see ***2.3 PV Module/array Optimum Tilt and Orientation Angles***). As stated earlier, there are formulas that can be used for calculating tilt angle for a specific day at specific latitude. The PV array orientation should always be due south in the northern hemisphere and due north in the southern hemisphere.

6.5.1.3 PV module supporting structures

PV module supporting structures for a standalone PV system installed on a building's roof should be compatible with the PV module frame and the building roof material (see ***4.2.3 PV module supporting structure***). Supporting structures for standalone PV modules are usually fixed on the ground or on poles that are fixed to the ground.

6.5.2 Battery for standalone PV system

The batteries that are generally used in standalone PV systems are lead acid deep-cycle batteries. The standalone PV system batteries are different from car batteries; they are designed to have a longer life cycle under daily battery discharge and recharge (known as battery cycle), whereas the car batteries are designed to provide a high current for short periods (starting-current) and have a shorter life cycle. The number of times the battery can be charged and be discharged determines the battery life (number of cycles).

Here, we will be concerned with lead acid batteries used in standalone PV systems. These batteries may be divided into two groups, sealed and unsealed. All sealed batteries are valve regulated in order to allow gas to escape during the charging and discharging of the battery. Sealed or valve regulated batteries come in three sub-types, wet, AGM (absorbed glass matt), or gel (gel electrolyte). The electrolyte in a sealed battery cannot be topped up regularly. The sealed type are maintenance free, and most sealed batteries are generally referred to as shallow cycle batteries because most sealed batteries have low allowable maximum depth of discharge (DOD). The allowable maximum depth of discharge is the maximum percentage of battery capacity that a battery can discharge without possible damage to itself. The DOD of some shallow batteries can be as low as 20% of battery capacity. The unsealed type (flooded cell) requires maintenance, and the electrolyte should be topped up regularly with distilled water. Flooded cell batteries have a high allowable DOD, some as high as 80% of battery capacity, and they are generally referred to as deep-cycle batteries. Note that the flooded cell lead acid battery may be sealed or valve regulated, which makes it maintenance free. The valve regulated flooded cell battery usually has a small nominal capacity. The batteries in each of the two types of batteries, i.e. the maintenance-free batteries (sealed or valve regulated) and the batteries that require maintenance (unsealed), can also defer because of the way they are manufactured. For example, the electrodes of some batteries can be made up of plates of either lead-calcium alloy or lead-antimony alloy. A flooded cell lead acid battery with electrodes made from lead-antimony alloy will lose more water from the electrolyte (dilute sulphuric acid) than a similar battery with lead-calcium electrodes.

Battery plates may be tubular or flat. Some batteries have thicker plates than others. Batteries with thicker plates are heavier and will generally last longer and have greater capacity (amp-hour) compared to batteries with smaller plates. Likewise, batteries with tubular plates, usually the 2V flooded cell batteries will last longer and have greater capacity (amp-hour) than batteries with flat plates, usually the 12V and 6V batteries. The 2V flooded cell battery has a life cycle of 20 to 25 years. Good quality 12V and 6V batteries have a life cycle of 10 to 12 years.

The usual battery voltage of gel cell batteries is 12V, while that of the flooded cell is usually 2V or 6V. The sealed batteries, for example the gel cell type batteries, are often used in small standalone PV systems while the flooded cell batteries are often used in larger standalone PV systems.

The nickel-cadmium battery is also used in standalone PV systems. It is a deep cycle, maintenance-free battery. The nickel-cadmium battery can be fully discharged without getting damaged and it can withstand temperatures below zero (the electrolyte does not freeze). The nickel-cadmium battery has nickel and cadmium as electrodes and a solution of potassium hydroxide (alkaline solution) and water as electrolyte.

The lead acid battery is made up of 2V electrochemical cells, connected in series, in order to achieve a desired battery voltage. This means that a 12V battery has six cells connected in series (2V × 6 = 12V). When fully charged, at nominal specified cell temperature, usually 20^0C (consult the battery's technical data sheet or the battery manufacturer), a battery cell has a voltage of about 2.12V and when fully discharged has a voltage of about 1.96V. The charging voltage should be greater than the battery's nominal voltage in order to fully charge the battery. Unlike the nickel-cadmium battery, lead acid batteries should never be fully discharged. In fact, the maximum depth of discharge of the lead acid battery must never be exceeded; doing so can reduce the battery's life cycle or completely damage it. Care should also be taken to never overcharge sealed batteries, including nickel-cadmium batteries. When a battery is being charged the electrolyte heats up and gas is released, and over-charging the battery can cause it to explode, with catastrophic results.

Lead acid batteries used in standalone PV systems are usually 2V, 6V or 12V batteries, as indicated above, and each type (same battery type with same voltage) can be connected in series and or parallel to achieve required PV system voltage and current, respectively, and forming what is known as battery bank (see **Figs. 21**, **22** and **23**). PV system voltage is usually 12V, 24V, or 48V for larger PV systems. The battery voltage should be within the inverter input voltage range, and the battery current should not be greater than the maximum DC input current of the inverter.

The standalone PV battery stores electrical energy, which is consumed by the load during the night or during cloudy days, as stated earlier. During the day, when the sky is clear, most of the electrical energy from the PV module/array is consumed by the load, and the excess energy is used for charging the battery (see ***6.5.3 Battery charge controller***). The battery also stabilises the voltage and current by eliminating transients and it also provides surge currents, e.g. starting currents for fans or pumps.

6.5.2.1 Battery capacity

The rated capacity of a battery for standalone PV systems is given in amp-hour (Ah), at a certain number of hours for discharge (C_{RATE}). The rated capacity of a battery refers to the amount of electrical energy the battery is capable of delivering once under favourable conditions, i.e. under the specified temperature and the specified rate of discharge. The battery capacity can vary due to change of the discharge rate, change of temperature, age, and type of battery. The discharge rate is the ratio of the battery rated (nominal) capacity and the discharge current. For example, if the battery rating is 100 Ah at a discharge rate of 10 hours (100 Ah at C_{10}), the discharge current would be 10 A. Consider the same battery with a discharge current of 15 A, the battery may probably be discharged in only six hours. In this case, the battery capacity would be 90 Ah at C_6. Therefore, the nominal capacity (Ah) of a battery will decrease with an increase of discharge current, and vice versa. This means that if the discharge current is increased the battery capacity will decrease. The battery capacity is also affected by battery temperature, and this capacity will decrease by 1% with a battery temperature decrease of 1^0C, and vice versa. The battery's technical data sheet should show the battery's nominal operating temperature; if not, the battery manufacturer should be consulted. For example, if the battery temperature has dropped 10^0C below the rated temperature of say 20^0C, when the battery is fully charged its capacity will only be 90% of its rated capacity. This means that the battery charging voltage is increased by the decrease of battery temperature, leading to reduced battery capacity and vice versa. Some charge controllers have means of sensing the battery temperature and then adjusting the charging voltage accordingly, so that the battery is charged to its rated capacity. The flooded cell battery's temperature should not be allowed to fall too low, to avoid the battery electrolyte being frozen (should the temperature fall below zero), which could damage the battery. When the battery is fully discharged the specific density of the electrolyte is at its minimum, i.e. at about 1.1 g/cm^3 at rated battery temperature (check the battery's technical data). This means that most of the sulphur would have combined with the lead electrodes leaving a more diluted electrolyte (made up of much more water than the sulphuric acid). Therefore, the freezing point of the electrolyte would be at about 0^0C. When a battery is fully charged the specific density of the electrolyte is at its maximum, i.e. at about 1.28 g/cm^3 at rated battery temperature, and the freezing point of the electrolyte is at lowest (check the battery's technical specification). Note that the specific density (1.28 g/cm^3 maximum and 1.1 g/cm^3 minimum) of the electrolyte of a lead acid battery in warmer regions, where the temperature is always above zero, may be a little less (consult the battery data sheet or the battery manufacturer). In general, a lead acid battery's temperature should not be allowed to fall below -10^0C.

6.5.2.2 Autonomous days and system availability

In a standalone PV system the battery/battery bank capacity should also take into account the amount of energy required during days when there is less or no sunshine, i.e. during cloudy weather.

The days during which there is less or no sunshine, due to cloud cover, are referred to as autonomous days or days of autonomy. Therefore, the size (Ah) of the battery/battery bank, as obtained from the PV system load calculation, must be multiplied by the number of days of autonomy and divided by the product of temperature factor and the factor for maximum depth of discharge (DOD). The days of autonomy can be calculated using a standard formula (see ***5.3 Example 6***).

6.5.2.3 Usable battery capacity

The usable battery capacity is less than the rated battery capacity. This is because the battery can only be discharged up to a certain level of discharge. The usable battery capacity is the amount of energy that is available from the battery to the load/s, and it is determined as follows:

$$\text{Usable battery capacity} = \text{rated battery capacity} \times \text{battery maximum DOD}$$

6.5.2.4 Batteries in Series

When batteries are connected in series the voltage of the battery bank is equal to the voltage of one battery multiplied by the number of batteries connected in series, while the amp-hour (Ah) remains the same as for one battery. **Fig. 21** below shows voltage (V), amp-hour (Ah), and energy (W-h) for four batteries connected in series.

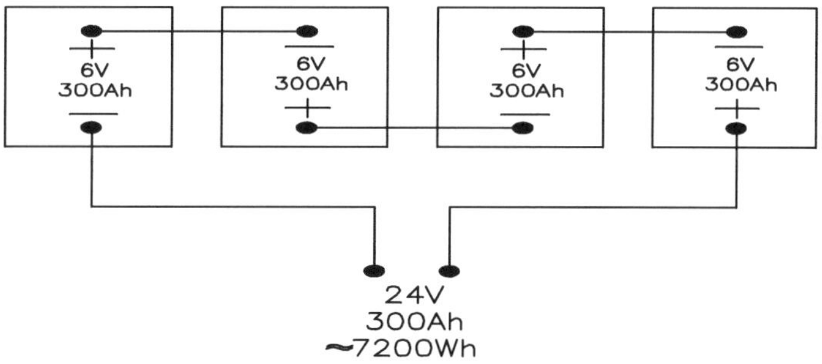

Figure 21: Batteries connected in series

6.5.2.5 Batteries in parallel

The number of batteries connected in parallel should preferably be less than three. This is done to mitigate chance of unequal charging between one battery or battery string and the next. When batteries are connected in parallel the amp-hour of the battery bank is equal to the sum of the amp-hours of batteries connected in parallel, while the voltage of the battery bank remains the same as for one battery. **Fig. 22** below shows voltage (V), amp-hour (Ah), and energy (W-h) for batteries connected in parallel.

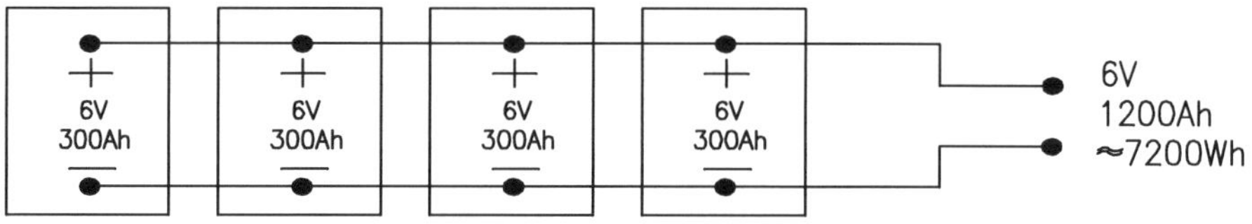

Figure 22: Batteries connected in parallel

6.5.2.6 Batteries in series and parallel

When batteries are connected in series and parallel the amp-hour of the battery bank is equal to the sum of the amp-hours of the battery strings in parallel, while the battery bank voltage becomes the voltage of one battery multiplied by the number of batteries in series (in a battery string). **Fig. 23** below shows voltage (V), amp-hour (Ah), and energy (W-h) for batteries connected in series and parallel.

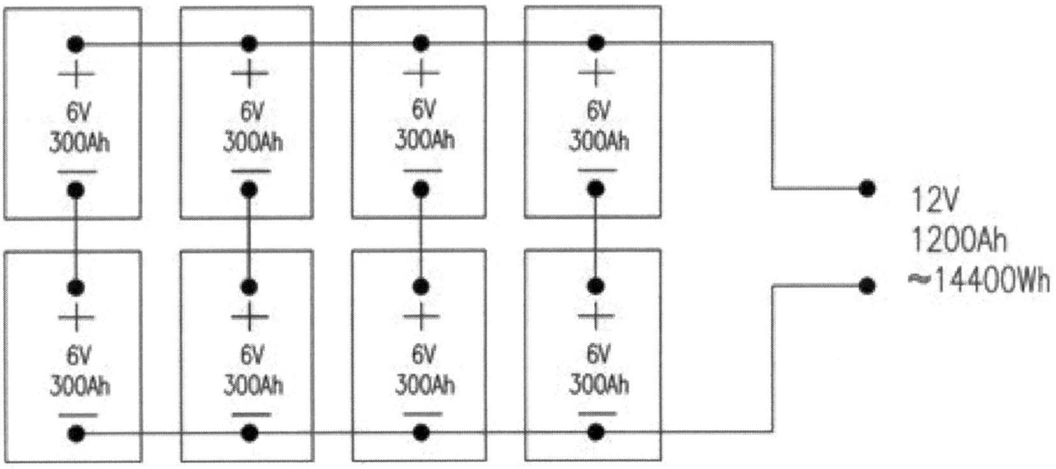

Figure 23: Batteries connected in series and parallel

6.5.2.7 Battery life cycle

The number of times the battery can be fully charged and fully discharged (to battery maximum DOD) before the battery can no longer be recharged to its rated capacity (but to only about 80% of rated capacity) will determine the battery's life cycle. The battery's life cycle is the number of times the battery can be discharged and re-charged. The different types of batteries have different life cycles. The deep-cycle flooded cell battery has the longest life cycle compared to other battery types used in standalone PV systems. The flooded cell has a maximum DOD of 80% of rated capacity. This means that the battery can be discharged safely to 80% of its rated capacity (Ah) without the battery getting damaged. Lead acid batteries for standalone PV systems should not be discharged to below a certain specified DOD. If the battery is over-discharged it can easily be damaged. If a battery operates at a lower DOD regime than its maximum rated DOD, the battery's life cycle becomes longer. In fact, if a battery with 80% maximum DOD is discharged to only 50% of its rated capacity (Ah) its life cycle is almost doubled.

Each time during battery discharge sulphur from the electrolyte is deposited on the battery cell electrodes (negative electrode), forming lead sulphate, and each time during battery charge most of the sulphur, which had combined with the battery cell electrodes, is released back to the electrolyte, with a little remaining as lead sulphate. Additionally, if the battery is over-charged more frequently than is necessary the acid concentration of the electrolyte will cause corrosion of the positive electrode. This means that the battery will be slowly degrading, and the battery life cycle will be reduced. Therefore, it is necessary to charge the battery regularly, and in accordance with the battery's technical data sheet or manufacturer's instruction. If a flooded cell battery undergoes several battery cycles without being charged fully, stratification (of the electrolyte) occurs, i.e. layers of different acid concentration in the electrolyte are formed. There are three stages of charging regimes: the bulk charge (at a high current rate), the absorb charge (at a slightly reduced current), and the float charge (at a trickling current). Some charge controllers are so designed that they can provide charging regimes for different battery types, e.g. for gel cell and flooded cell. The flooded cell requires a bulk charge that mixes up the electrolyte and reduces stratification. Sealed batteries should not be over-charged, as explained earlier. When charging the battery, the manufacturer's instructions must always be strictly followed.

Although increasing the battery temperature increases the battery capacity (Ah), the battery should not be allowed to operate beyond temperature specified by the manufacturer (see ***6.5.2.1 Battery capacity***). If the battery is allowed to operate at higher temperatures than the battery rated temperature, the life cycle of the battery will be reduced and it could be damaged permanently. Therefore, it is always necessary to check for the battery's rated temperature in its technical data sheet or to ask for it from the manufacturer.

6.5.2.8 Selecting a battery for a standalone PV system

When selecting a battery for a standalone PV system the following, among other things, should be considered:

- Battery capacity: the battery capacity must be adequate to meet energy requirements during cloudy days.
- Efficiency: consideration should be made regarding the amount of energy fed into the battery and how much, in turn, is received from the battery.
- Voltage and current: ascertain whether the battery can be configured with the PV array and/or inverter (where inverter is needed) to deliver the required voltage and current.
- Physical size and storage: the physical size of the battery and available space for it should be considered.
- Maintenance: decide whether to use a maintenance-free battery or a battery that requires regular maintenance.
- Availability: the battery should be easily available on the market.
- Cost: the price of the battery, including transport costs, should be affordable.
- National codes and regulations: the battery should meet national codes and regulations.

6.5.2.9 Battery location (housing)

The battery should be installed in a battery housing or container, and protected against extreme temperatures and water.

The battery housing or container should be properly ventilated so as to maintain the battery's rated temperature, and to allow the hydrogen gas, which is released by the battery during charging and discharging, to escape. Good ventilation should be capable of expelling the hydrogen gas from the battery housing (room). Hydrogen is flammable and can cause explosions. Therefore, naked fire and electric switches should be avoided in the battery housing (room).

The battery should not be placed directly on to a concrete floor because any spillage of acid (sulphuric acid) contained in the battery will produce a chemical reaction with cement materials. The battery must be placed on a non-corrosive battery rack or container. The battery housing and/or container should comply with the manufacturer's instructions.

6.5.3 Battery charge controller

The battery charge controller charges the battery/battery bank and protects it from over-charge and over-discharge. The battery charge controller senses the state of the battery charge. When the battery is fully charged the charge controller disconnects the supply to the battery (HVD – high-voltage-disconnect), and when the battery's state of charge falls below the acceptable level the charge controller disconnects the supply from the battery to the load (LVD – low-voltage-disconnect). For example, for a charge controller connected to a 12 VDC battery: The charge controller will disconnect the supply to the battery when the battery's voltage reaches 14 VDC and reconnect the supply to the battery when the battery's voltage falls to 12.8 VDC; the charge controller will disconnect the battery from the load when the battery's voltage falls below 11.5 VDC and reconnect the battery to the load when the battery's voltage reaches 12.4 VDC (consult the charge controller's technical data sheet). Some charge controllers have temperature compensation, which is very important where the battery's temperature varies more than ±5 ^{0}C from ambient temperature. This means that in colder temperatures the charging voltage can be increased by the charge controller in order to achieve the same battery state of charge, and vice versa (see ***6.5.2.1 Battery capacity***). Some charge controllers may also have three stages of charge control. This means that these charge controllers will provide bulk charge (at a high current rate), absorb charge (at a slightly reduced current), and float charge (at a trickling current). Some charge controllers may also provide maximum power point tracking of the PV array's power output.

The standard requires that the battery charge controller's rated current must not be less than 125% of the PV array short-circuit current. The correction factor takes into account the change in operating conditions of the PV system (PV module manufacturer's instructions, controller manufacturer's instructions, national codes and regulations).

The PV array's output voltage must be within the input voltage range of the battery's charge controller. If the input voltage range of the charge controller is exceeded, the controller would be damaged permanently. The charge controller should be installed near the battery, at most not more than two metres from the battery.

A charge controller should display some or all the following information (depending on size and/or cost):

- Model number
- Nominal system voltage
- Rated current
- Charging array current indication
- Charging voltage indication
- Low voltage warning indication
- Reverse current protection
- Equalisation charging voltage
- Boost charging voltage

- Float charging voltage
- Voltage activation set-points
- Temperature compensation
- Terminals
- Working temperature

6.5.3.1 Selecting battery charge controller

When selecting a battery charge controller for a standalone PV system the following, among other things, should be considered:

- Controller input voltage: the charge controller input voltage should be higher than the open-circuit voltage of the PV module, or higher than the open-circuit voltage of the PV string, where the PV array has two or more PV modules in a PV string. The controller voltage rating should be rate greater than the PV array / string voltage corrected for lowest site temperature.
- Controller rated current: the charge controller current should be equal to or greater than the total PV array short-circuit current multiplied by a correction factor (PV module manufacturer, national codes and regulations should be consulted).
- The charge controller should be compatible with the battery. Consult with the battery supplier or manufacturer when selecting the battery charge controller.
- Reverse current protection: some charge controllers have reverse current protection, and you may require one with this facility.
- Voltage activation set-point: the set-point at which the controller will disconnect the PV array when the battery is fully charged, the set-point at which the controller will reconnect the PV array when the battery voltage falls below a specified level (e.g. for a 12V battery the reconnect set-point is near 12.8V), the set-point at which the charge controller will disconnect the load from the battery when the battery voltage falls below a specified level, and the set-point at which the controller will reconnect the battery to the load.
- Charging regimes: sensing battery voltage, disconnecting the battery when the battery is fully charged, reconnecting the battery when the battery reaches a certain level of discharge in order to maintain battery charge, and to disconnect the battery when the battery reaches the low level of depth of discharge. Some charge controllers have different regimes for charging different types of batteries, e.g. flooded cell batteries may require a little over-charge to mix up the electrolyte to mitigate stratification, whereas gel batteries should never be over-charged.
- Temperature compensation: some charge controllers have temperature compensation, i.e. they have the ability of sensing the battery's temperature and adjusting the charging voltage accordingly, to enable the battery to be charged to its nominal capacity (essential in very cold regions).

- Maximum power point tracking: Some charge controllers are capable of tracking the maximum power point (MPP) of a PV module/array. You may require maximum power point tracking for your PV system.
- Cost: the price of battery charge controller, including transport costs, should be affordable.

6.5.4 Inverter for standalone PV system

All standalone PV systems that are required to supply power to AC loads have inverters. The inverters for standalone PV systems are different from inverters for grid-tied PV systems, and they must not be interchanged with grid-tied inverters. For example, a grid-tied inverter must have a pure sine wave and mechanism for synchronisation (with the utility grid power supply). An inverter for a standalone PV system may have a square wave or modified wave, or pure sine wave, but it does not have a mechanism for synchronisation. The selection of inverter for standalone PV system, with regard to waveform, will depend on the type of waveform required by the electrical load. Most electrical load will require a sine wave. However, some electrical load may require a square or modified sine wave: a square wave is suitable for resistive loads and incandescent lamps; a modified sine wave is suitable for lights, electronic equipment, and most motors.

6.5.4.1 Requirement of inverters for standalone PV systems; system voltage and peak current

The DC input voltage of the inverter should determine the standalone PV system voltage (which should be the battery/battery bank voltage), while the inverter AC output power should determine the inverter's rated power. The rated power of the inverter will depend on the PV system's load requirement (see, ***6.5.4.2 Selection of inverter for standalone PV system***). The inverter's rated power should be about 125% to 130% of the connected load. The extra 25% to 30% is to cater for possible overload and future expansion.

PV system voltage:

The PV array voltage is determined by the battery or battery bank's voltage, which in turn is determined by the size of the inverter (where an inverter is required). The PV array's voltage should be sufficient for charging the battery/battery bank. This means that the PV array's voltage should not be less than the battery/battery bank's charging voltage. The battery's charging voltage is 120% of nominal battery/battery bank voltage, as a minimum (see ***6.5.1 PV module for standalone PV system*** and ***6.5.2 Battery for standalone PV system***). If the PV array is used for charging the battery, the PV system load should be disconnected during charging.

The decision on the PV system voltage will depend largely on the voltage of most loads and the total current. In fact, the PV system voltage should be equal to the voltage of the load with the largest circuit current. The other loads may be directly connected to the PV system's output if the PV system voltage is equal to the voltage of the largest load. This means that any load that has a voltage that is different from the PV system's voltage will be connected via a DC-DC converter.

The selected system voltage should be such that the cable sizes are minimised, while keeping the voltage drop at an acceptable level and in accordance with national codes and regulations. As a rule of thumb, the current of the largest circuit must not exceed 100A and the current of the other circuits must be less than 20A. Where an inverter is connected, the inverter input voltage should determine the PV system voltage. The inverter should be capable of meeting the total AC load, while keeping the DC circuit current less than 100A (rule of thumb). For example, if the single phase required power at the inverter terminals is 2000W, disregarding losses in the inverter circuits the DC power at the inverter input terminals would also be 2000W. And if the DC voltage was 12V, the DC current to the inverter (the main circuit current) would be 2000W ÷ 12V = 166.67A. Since in one circuit the DC current should not exceed 100A, we must increase the inverter DC input voltage. With 24 DC input voltage, the DC current would be 2000W ÷ 24V = 83.33A. The current of the DC circuit to the inverter should never exceed the maximum DC input current of the inverter. The 24V DC is recommended and is acceptable because the main circuit current (current to the inverter) is below 100A. Other source currents to DC loads should be limited to no more than 20A. This helps to reduce equipment sizes and DC cable sizes, while keeping voltage drop at a minimum.

Peak current:

The PV system's peak current is equal to the total load of the PV system divided by the PV system's voltage. The peak current is used for calculating the cable size between the battery/battery bank and the battery charge controller. However, the cable size between the battery and the controller is usually provided by the battery manufacturer or supplier.

6.5.4.2 Selecting inverter for standalone PV system, and inverter label information

When selecting an inverter for a standalone PV system the following, among other things, should be considered:

- The inverter's DC input voltage should be such that the variation of the battery voltage remains within the inverter voltage range, that is to say, the battery voltage must not exceed the higher limit of the inverter voltage range and must not fall below the lower limit of the inverter voltage range. Increasing the input voltage of the inverter (which determines the PV system voltage) will lead to reduced DC cable sizes, while maintaining voltage drop at an acceptable level. The inverter output voltage must be within 10% of inverter output voltage, i.e. 10% of 230V.
- The inverter should have a stable frequency (50Hz). The deviation from 50Hz should not be more than 5%.
- The inverter should be capable of sensing the input over-voltage or under-voltage and cut off the circuit. This is to protect the inverter from possible damage.
- Protection against over-current (150% of rated current).
- Protection against short-circuiting of input and output terminals (not exposed to contact).
- Protection against reversed polarity of input and output terminals.

- Duty rating of the inverter. See whether the inverter can supply: rated power without overheating for four hours at ambient temperature of 25^0C; 125% of its rated power for one minute safely; and 150% of its rated power for two seconds safely.
- The inverter should have good efficiency at full and part load.
- The output waveform must be suitable for the connected load. The output waveform may be square, modified, or pure sine wave. Most selected inverters have pure sine wave, since most AC loads require this.
- Inverter standby current: this current should be low, and not exceed 3% of the inverter's rated current.
- The inverter should be suitable for the environment in which it is required to operate.
- The inverter should meet national codes and regulations.

***Inverter label information*:**

Each standalone PV inverter should have a label with minimum information as follows:

- Manufacturer;
- Model number;
- Continuous power (W);
- Waveform: square, modified, or pure sine wave, and
- If pure sine wave, harmonic distortion percentage;
- Frequency (Hz);
- DC system voltage (V) ±V;
- AC system voltage (V) ±V;
- Efficiency at 25% of rated power;
- Efficiency at rated capacity;
- Low voltage protection;
- Over-voltage protection;
- Over-current protection;
- Short-circuit protection;
- Lightning protection.

6.5.4.3 Inverter location

The inverter should be installed as near as possible to the PV array or PV array junction box so that the PV string cables and main DC cables are kept as short as possible

The inverter should be installed where it will not create disturbing noise (noise pollution) to the people living in the building. The inverter should be able to operate within its rated temperature.

Excessive inverter temperature leads to reduced inverter efficiency. The inverter should also be protected against dust and humidity. In fact, the inverter location must comply with the instructions of the inverter manufacturer.

6.6. Other Components for Standalone PV System

6.6.1 Cables

The DC cables must be UV resistant and suitable for PV system (see ***4.2 Other Components of Grid-Tied PV System***). Cables installed in the battery room should be suitable for or protected against the corrosive environment due to the battery electrolyte. The size of the cable from the charge controller to the battery/battery bank is usually provided by the battery manufacturer or supplier. The voltage drop to the battery should not exceed 2% and the voltage drop from battery to DC loads should not exceed 5%. (You check national codes and regulations concerning permitted voltage drop.) Calculations of cable sizes and selection of equipment must be carried out in accordance with national codes and regulation.

6.6.2 Isolators, fuses, switched fuses, circuit breakers

Isolators, fuses, switched fuses, and circuit breakers, on the DC circuit, should be suitable for DC operation (see ***4.2.5 Fuses, switches (isolators, connectors, circuit breakers, switch fuses)***).

6.7 Standalone PV system Design

6.7.1 Steps in the design of standalone PV system

- Initial discussion with client (owner)
- Site visit
- Preliminary design and an estimate of the PV system energy output
- Design

6.7.1.1 Initial discussion with client (feasibility assessment)

Discuss energy serving methods with the owner; for example, find out which energy efficient appliances can be connected to the PV system (see ***5.1.1 Initial discussion with client (owner)***). This can help in reducing the total load, while maintaining the required electrical energy. Larger appliances can be replaced with units with less wattage. The use of DC lights such as DC compact florescent lamps, and DC appliances, such as DC refrigerators or DC fans, should be considered instead of their AC counterparts. If the appliance is, for example, an AC refrigerator it can be replaced with a DC unit where a DC unit is available. Alternatively, the refrigerator can be removed from the PV system, and the refrigerator can be gas-powered instead. The AC load will be connected to the inverter. Incandescent lamps should be avoided; instead, the use of fluorescent lamps, which consume less energy, is recommended. In fact, connection to the PV system of resistive loads, such as toasters, cooking stoves, and water heating appliances, should be avoided. For cooking, a gas stove or wood-fired stove could be used instead of solar power. For water heating, a direct solar water heating system could be used. Heavy loads such as a washing machine or drier should be left out, unless a PV hybrid can be considered. The wattage of each lamp or appliance should be determined. The wattage can always be found on the lamp and on the appliance label or in the manual of the appliance. The time during which each light or appliance will be in use is estimated and agreed upon, and the number of days each light or appliance will be switched on is decided and agreed upon. The period during which the PV system is most required – for example, during summer, winter, or the whole year – should be decided. The system availability should also be taken into account.

6.7.1.2 Site visit

Find out the site's latitude, longitude, wind speed, and height (altitude). Determine solar radiation (KW /m^2) (see ***6.3 Solar Resource for Standalone PV System***) and the extreme temperatures at the site. Check the available space for the PV array, to see whether it is suitable and adequate for the installation of the PV array. Estimate PV modules' inter-row spacing. The available space should be free of any shading (see ***5.1.2 Site visit***). The PV modules/array for a standalone PV system should always face true south for locations in the northern hemisphere and true north for locations in the southern hemisphere. The PV module tilt angle must be the PV module optimum tilt angle for the design month at the location site. Find the solar irradiation (kWh / m2 / day) at the site location. The irradiation obtained should be on an optimally tilted and oriented PV array. If the irradiation obtained is on a horizontal surface, the irradiation should be adjusted for an optimally tilted PV array. Decide on the position of the inverter (to be agreed with the client), PV junction box, battery, and battery charge controller. Decide also on DC cable routes and estimate the cable lengths.

6.7.1.3 Preliminary design and electrical energy estimate

Estimate the total PV system load. This is the sum of the DC loads plus the sum of the AC loads – the sum of all the lights and appliances connected to the PV system. The total wattage of all the lights and appliances selected during the initial discussion with the client is assessed. It may be necessary to divide the load in order to come up with two or more PV systems. If one PV system breaks down the other system/s will remain operating. Decide on the system's voltage and the size of the inverter. The standard requires that if the PV system size is 1.5 kW to 2.5 kW a system voltage of 24 VDC is recommended, and below 1.5 kW and above 2.5 to 5kW will require a system voltage of 12 VDC and 48 VDC, respectively (see ***6.1.2 PV hybrid system***). The PV array power to load ratio will be determined at a later stage. If the PV array power to load ratio is greater than 0.3 a PV hybrid system may be considered (see ***6.1.2 PV hybrid system***). Using the total load to be connected to the inverter, the inverter is selected (see ***6.5.4.1 Requirement of***

inverters for standalone PV systems and ***6.5.4.2 Selecting inverter for standalone PV system, and inverter label information***). Estimate the electrical energy (W-h) required by each load (both DC and AC loads) and calculate the total. This is the total energy (W-h) consumed. Select the battery type. Estimate the energy produced by the PV system. This is the total watt-hours, for both DC and AC load, divided by the product of the battery and wiring efficiency factors. This is in line with the design method used in ***6.7.2 Example 8*** (see ***6.4.1 Estimating PV system energy output for standalone PV system***). Make a sketch plan of the standalone PV system. The PV array may be divided into two or more PV arrays, each array connected to a separate battery charge controller, and the charge controllers connected in parallel to the battery/battery bank.

6.7.1.4 Design stage:

System loads:

- Calculate the total power (W) for the DC loads;
- Calculate the total power (W) for the AC loads;
- Calculate the total power (W) for both DC and AC loads;
- Calculate the PV system's peak current;
- Decide on the duty cycle (number of hours the load will be connected and for how many days per week) for each load;
- Calculate the energy (W-h) consumed each day by the DC loads (taking into account the efficiency of each load);
- Calculate the energy (watt-hour) consumed each day by the AC loads (taking into account the efficiency of each load);
- Calculate the total energy (W-h) consumed by both the DC and AC loads each day.

Decide on the PV system voltage:

- Select inverter and PV system voltage.

Amp-hours consumed and required to be produced each day:

- Calculate the total energy (Ah), for DC and AC loads, required for consumption each day;
- Calculate the energy (Ah) produced each day;

Select battery type, and:

- Determine the needed battery capacity for the PV system;
- Select the size of battery (rated Ah);
- Find the number of batteries that can be connected in series and in parallel;
- Calculate the PV system's nominal battery capacity (Ah)
- Determine the usable battery (battery bank) capacity.

Determine the design current at a specific PV array tilt angle, and:

- Select the PV module and determine the PV array size (PV modules connected in series and in parallel);
- Determine the size of the controller and select type of controller;
- Determine the sizes of cables' isolation switches, protective devices, and other system components.
- Draw the final system layout.

6.7.2 Example 8

From discussion with the client:

Design a standalone PV system for supplying electricity to a house in Luanda, Angola and produce a standalone PV system layout (no cable or switching equipment sizing). The PV system will be required to provide power to the following:

- Lighting
- Television
- Radio
- Computer
- Printer
- Cell phone
- Microwave
- Refrigerator and freezer
- Fan

Lights and the refrigerator and freezer will be fed from the DC circuit, while the television, radio, computer, printer, cell phone, microwave and fan will be fed from the AC circuit.

Site inspection:

The site location is at 8.82^0 south 13.22^0 East and 42m high. Minimum and maximum temperatures at the site are 15^0 and 40^0, respectively. The PV module temperature is expected to range from 15^0 to 70^0. There will be no partial shading of the PV array. The PV modules will be mounted on poles fixed on to the ground. The PV module inter-row spacing is estimated as 1.2 metres. The insolation data (sun-hours) at the site is shown in **Table 5** below.

Table 5: Solar irradiation at the site: latitude 8.82^0 south and longitude 13.22^0 East

	Jan Feb Mar	Apr May Jun	July Aug Sept	Oct Nov Dec	YR
Latitude Tilt -15(^{0}) Fixed Array	5.92 6.07 5.43	4.89 4.60 4.18	3.36 3.70 4.57	5.06 5.60 6.16	4.96
1-Axis Tracking	7.62 7.83 7.02	6.19 5.61 5.01	4.17 4.75 5.96	6.66 7.27 7.87	6.33
Latitude Tilt (^{0}) Fixed Array	5.56 5.87 5.49	5.19 5.11 4.75	3.713.954.68	4.97 5.31 5.72	5.03
1-Axis Tracking	7.20 7.66 7.19	6.686.34 5.80	4.78 5.21 6.21	6.60 6.93 7.36	6.50
Latitude tilt +15(^{0}) Fixed Array	4.94 5.40 5.30	5.27 5.42 5.14	3.93 4.04 4.60	4.664.77 5.02	4.87
1-Axis Tracking	6.28 6.96 6.89	6.76 6.70 6.27	5.11 5.36 6.08	6.11 6.11 6.33	6.25
Two-Axis Tracking	7.67 7.84 7.20	6.796.73 6.34	5.14 5.37 6.23	6.69 7.30 7.95	6.77

Note: This table is for this example only; check the table's accuracy before using it for any other purpose.

The PV array shall be mounted on poles, north of the house and seven metres from it; the junction box shall be installed (on the PV array supporting frame) below the PV array. The battery/battery bank shall be installed in a purpose built battery room at the back of and adjacent to the garage, which is attached to the main building. The control centre and the inverter shall be installed in the garage that is adjacent to the house. This arrangement will reduce DC and AC cables to minimum lengths. The DC and AC boards shall be installed in the kitchen (see **Fig. 24** below).

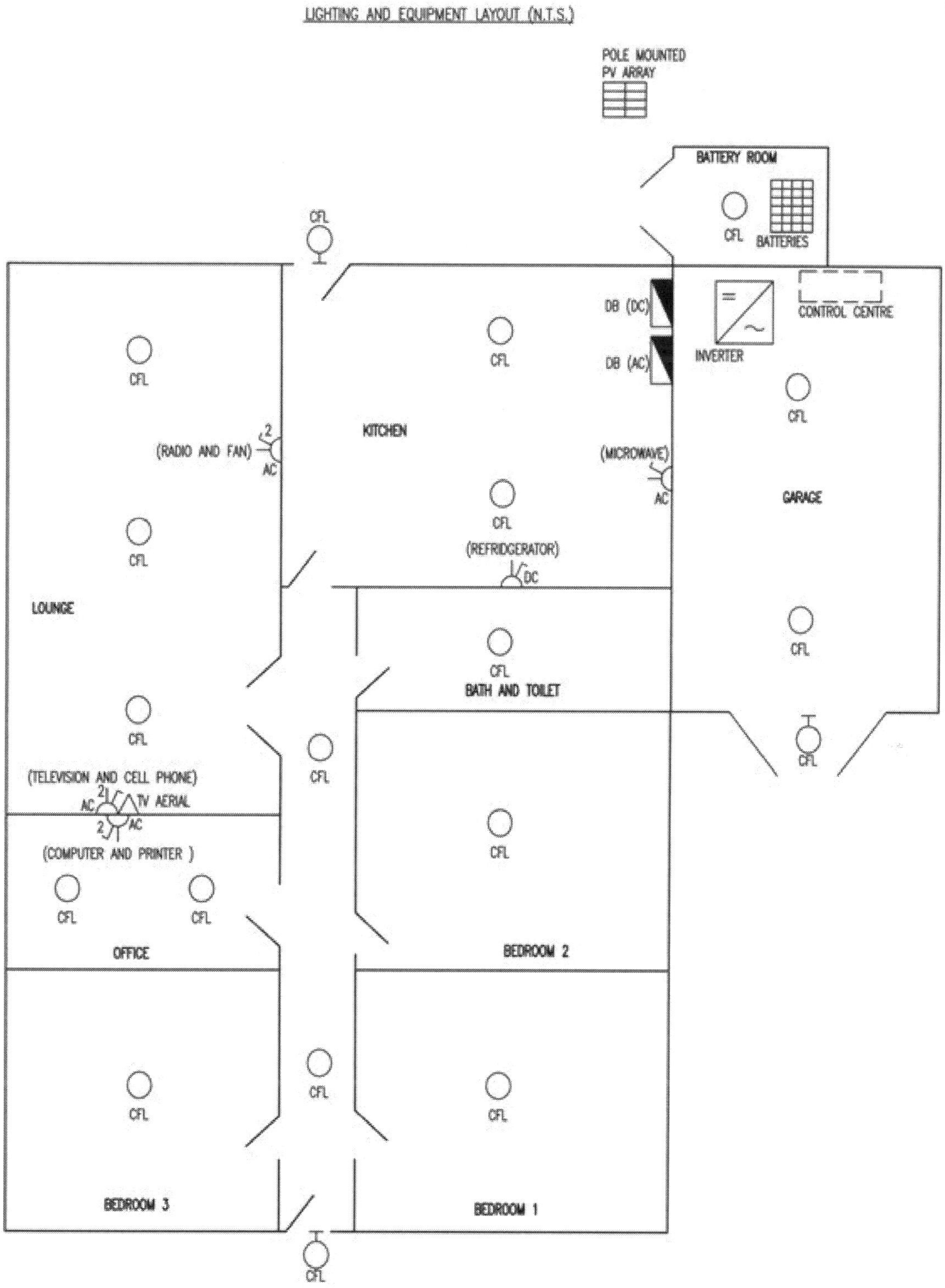

Figure 24: Lighting and equipment layout

Battery Room:

The battery room shall be properly ventilated to allow hydrogen gas released by batteries during charging and discharging to escape and to allow batteries to operate within the allowable battery temperature. The ventilation shall be carried out in accordance with the manufacturer's or supplier's instructions. The batteries shall be installed on battery racks and in accordance with the manufacturer's or supplier's instructions. The light fitting (luminaire), in the battery room, shall be water tight, non-corrosive, and explosion proof with flameproof construction. The switch for the luminaire shall be installed outside the battery room and be weatherproof.

Determination of PV System Load:

DC Loads:

Appliance	Location	Quantity	Power Rating	Total Load	Efficiency (η)	Duty Cycle (W-h / day)
CFL (24VDC)	3 bedrooms	3 ×	11W =	33W ÷	1 ×	1 × Lights

Days/Week	Watt-hours Consumed Per day
7/÷7 =	33

Appliance	Location	Quantity	Power Rating	Total Load	Efficiency (η)	Duty Cycle (W-h / day)
CFL (24VDC) Lights	Kitchen	2 ×	15W =	30W ÷	1 ×	4 ×

Days/Week	Watt-hours Consumer Per day
7/÷7 =	120

Appliance	Location	Quantity	Power Rating	Total Load	Efficiency (η)	Duty Cycle (W-h / day)

CFL (24VDC) Lights	Lounge	3 ×	11W =	33W ÷	1 ×	5 ×
Days/Week	Watt-hours Consumed Per day					
7/÷7 =	165					
Appliance	Location	Quantity	Power Rating	Total Load	Efficiency (η)	Duty Cycle (W-h / day)
CFL (24VDC) Lights	Office	2 ×	11W =	22W ÷	1 ×	1.5 ×
Days/Week	Watt-hours Consumed Per day					
7/÷7 =	33					
Appliance	Location	Quantity	Power Rating	Total Load	Efficiency (η)	Duty Cycle (W-h / day)
CFL (24VDC) Lights	Bath / toilet	1 ×	15W =	15W ÷	1 ×	1.5 ×
Days/Week	Watt-hours Consumed					

	Per day
7/÷7 =	22.5

Appliance	Location	Quantity	Power Rating	Total Load	Efficiency (η)	Duty Cycle (W-h / day)
CFL (24VDC) Lights	Passage	2 ×	9W =	18W ÷	1 ×	1 ×

Days/Week	Watt-hours Consumed Per day
7/÷7 =	18

Appliance	Location	Quantity	Power Rating	Total Load	Efficiency (η)	Duty Cycle (W-h / day)
CFL (24VDC) Lights	External	2 ×	11W =	22W ÷	1 ×	1 ×

Days/Week	Watt-hours Consumed Per day
7/÷7 =	22

Appliance	Location	Quantity	Power Rating	Total Load	Efficiency (η)	Duty Cycle (W-h / day) (for summer)
Refrigerator	Kitchen	1 ×	60W =	60W ÷	1 ×	9

Days/Week	Watt-hours Consumed Per day
7/÷7 =	540

Appliance	Location	Quantity	Power Rating	Total Load	Efficiency (η)	Duty Cycle (W-h / day) (for winter)
Freezer	Kitchen	1 ×	60W =	60W ÷	1 ×	5

Days/Week	Watt-hours Consumed Per day
7/÷7 =	300

Appliance	Location	Quantity	Power Rating	Total Load	Efficiency (η)	Duty Cycle (W-h / day)
CFL (24VDC) Lights	Garage	3 ×	11W =	33W ÷	1 ×	1

Days/Week	Watt-hours Consumed Per day
7/÷7 =	33

Appliance	Location	Quantity	Power Rating	Total Load	Efficiency (η)	Duty Cycle (W-h / day)
CFL (24VDC) Lights	Battery Room	1 ×	15W =	15W ÷	1 ×	1

Days/Week	Watt-hours Consumed Per day
1/÷7 =	2.14

Single Phase AC Loads:

Appliance	Location	Quantity	Power Rating	Total Load	Efficiency (η)	Duty Cycle (W-h / day)
Television (230V AC)	Lounge	1 ×	150W =	150W ÷	0.85 ×	4 ×

Days/Week	Watt-hours Consumed Per day

7/÷7 =	705.88

Appliance	Location	Quantity	Power Rating	Total Load	Efficiency (η)	Duty Cycle (W-h / day)
Radio (230V AC)	Lounge	1 ×	12W =	12W ÷	0.85 ×	**2** ×

Days/Week	Watt-hours Consumed Each day
7/÷7 =	**28.24**

Appliance	Location	Quantity	Power Rating	Total Load	Efficiency (η)	Duty Cycle (W-h / day)
Computer (230V AC)	Office	1 ×	300W =	300W ÷	0.75 ×	1.5 ×

Days/Week	Watt-hours Consumed Per day
7/÷7 =	600

Appliance	Location	Quantity	Power Rating	Total Load	Efficiency (η)	Duty Cycle (W-h / day)
Cell-phone	Office	3 ×	5W =	15W ÷	0.75 ×	0.5 ×

Days/Week	Watt-hours Consumed Per day
4/÷7 =	5.71

Appliance	Location	Quantity	Power Rating	Total Load	Efficiency (η)	Duty Cycle (W-h / day)
Printer (230V AC)	Office	1 ×	20 W =	20 W ÷	0.80 ×	0.1 ×

Days/Week	Watt-hours Consumed Per day
3/÷7 =	1.07

Appliance	Location	Quantity	Power Rating	Total Load	Efficiency (η)	Duty Cycle (W-h / day)
Microwave (230V AC)	Kitchen	1 ×	550W =	550W ÷	0.85 ×	0.2 ×

Days/Week	Watt-hours Consumed

	Per day
7/÷7 =	129.4

Appliance	Location	Quantity	Power Rating	Total Load	Efficiency (η)	Duty Cycle (hours / day)
Fan (230V AC)	Lounge	1 ×	40W =	40W ÷	0.80 ×	0.5

Days/Week	Watt-hours Consumed Per day
7/÷7	25

PV system total DC load:

The PV system total DC load (W) is equal to the sum of connected loads (W). So,

Total DC load (W) = (33+ 30 + 33 + 22 + 15 + 18 + 22 + 60 + 60 + 33 + 15) W = 341W

PV system total AC load:

The total AC load (W) is equal to the sum of connected AC loads. So,

Total AC load (W) = (150 + 12 + 300 + 15 + 20 + 550 + 40) W = 1087W

Disregarding the inverter losses, the power (W) at the inverter input terminals would be equal to 1087W.

This means that the PV system's total DC load (W) is equal to the total of the connected DC loads plus the total of the connected AC loads. So,

Total load (W) = 341W + 1087W = 1428W. This power will be used for determining the PV system's peak current.

Total watt-hours per day (consumed):

The total watt-hours consumed each day is equal to the sum of energy per day (W-h / day) used by the connected load. So,

Total watt-hours per day = (33 + 120 + 165 + 33 + 22.5 + 18 + 22 + 540 + 300 + 33 + 2.14 + 705.88 + 28.24 + 600 + 5.71 + 1.07 + 129.4 + 25) W-h / day = 2783.94 W-h / day.

Inverter selection and decision on PV system voltage:

The inverter determines the PV system voltage. The current to the inverter, which is the current of the largest circuit, is equal to the power (W) appearing at the inverter input terminals divided by the PV system voltage (VDC). Disregarding inverter losses, the power at the inverter output terminals (1087W) is equal to the power at the inverter input terminals. So:

a) If 12 VDC is selected as the PV system voltage, the source circuit current to the inverter (the largest circuit to load) and the source circuit current to other DC loads would be as follows:

i) Circuit current (to inverter) = 1087W ÷ 12V = 90.58A. This is acceptable because the source circuit current of the largest circuit is less than 100A.

ii) Circuit current (to other loads) = 341W ÷ 12V = 28.42A. This is not acceptable. The standard is that the source circuit to the smaller load should be kept to no more than 20A. However, this load may be supplied via a 24V DC-DC converter, which means that the circuit current would be 14.21A.

b) If 24 VDC is selected as the PV system voltage, the circuit current to the inverter (the largest circuit to load) and the source circuit current to other DC loads would be as follows:

i) Circuit current (to inverter) = 1087W ÷ 24V = 45.29A. This is acceptable because the source circuit current of the largest circuit is less than 100A.

ii) Circuit current (to other loads) = 341W ÷ 24V = 14.21A. This is acceptable because the smaller source circuit current is below 20A.

An inverter, which has an input voltage of 24VDC, is selected (see ***Selecting Inverter for Standalone PV System***). The selected PV system voltage of 24 VDC will result in reduced DC cable sizes and improved PV system efficiency.

iii) The rated inverter power will be 130% of total load power (1087W × 130% = 1413.1W) to make sure that the inverter will never be overloaded and to allow for future expansion. The selected inverter is rated at 1500W.

The specification of the selected inverter (rated 1500W) is as follows:

- Manufacturer: Sunshine Solar Limited
- Model: Sunshine Solar
- Serial No.: VP1500

Features:

- Soft start
- Over temperature protection
- Low battery voltage alarm
- Low battery voltage shutdown
- High battery voltage shutdown
- Overload protection
- Short circuit protection
- Polarity protection
- Polarity reverse connection protection

Specification:

- Rated power 1500W
- DC voltage (input voltage) 24V (21.6 – 31 VDC)
- DC current 75A
- AC voltage (output voltage) 230V (200V – 240V)

- Continuous AC power 1500W
- Waveform: pure sine wave (THD less than 3%)
- Surge capacity (W) 3000W; 2250W for 10 seconds; 3000W for 2 seconds
- Frequency 50Hz ± 0.1
- Efficiency 90% (full load) / 95% (1/3 load)
- No load consumption less than 0.4A
- Length 330mm
- Width 220mm
- Height 90mm
- Weight 2.5kg

PV system peak current:

The PV system's peak current is equal to the PV system's total load power divided by the PV system's DC voltage. So,

Peak Current = 1428W ÷ 24V = 59.5A

The PV system's peak current is used to determine the sizes of DC fuses and cables (between the battery and the battery charge controller). Usually, the size of the cable is provided by the battery manufacturer or supplier.

Total amp-hours consumed per day:

The amp-hours consumed each day are equal to the watt-hours consumed per day divided by the PV system voltage. So,

Total amp-hours per day = 2783.94 W-h / day ÷ 24V = 116 Ah / day.

Amp-hours produced per day:

The amount of electrical energy produced by PV array each day (Ah / day) should take into account cable efficiency and battery efficiency (including battery charge controller). The applicable efficiency factors are as follows:

Cable efficiency (η_{cab}) = 0.98

Battery efficiency (η_{BAT}) = 0.92

The amp-hours produced each day is equal to amp-hours consumed each day divided by the product of efficiency factors for cable and battery. So,

Amp-hours produced per day = 116 Ah / day ÷ (0.98 × 0.92) Ah / day = 128.66 Ah / day

Determining needed battery capacity:

Needed battery capacity is equal to the amp-hours produced each day multiplied by the number of days for battery storage (days of autonomy) divided by the product of the battery's maximum depth of discharge and the battery's de-rate temperature factors.

The maximum of the battery's depth of discharge is equal to 0.80 and the de-rate temperature is equal to 1 (obtainable from the technical data sheet or from the battery manufacturer); the battery storage days are approximately three days. The autonomy days have been determined from the following standard formula:

D = –48T + 4.58, where T are sun-hours for the design month. So,

D = –0.48 × 3.93 + 4.58 = 2.69 days. This means three days approximately.

Needed battery capacity = 128.66 amp-hours / day × 3 days ÷ 0.80 ÷ 1 = 482.48 Ah.

Battery selection:

A 550 Ah deep cycle lead acid flooded cell battery, 2V, lead-antimony acid battery, with 80% depth of discharge is selected.

Technical data of the selected battery:

Model:	OPzS Solar 550
Part number:	NVSL020550WCOFA
Voltage:	2V
Nominal capacity:	550 Ah
Length:	126mm
Width:	208mm
Height:	511mm
Installed length:	136mm
Weight incl. acid:	26.7 Kg
Weight of acid:	8.10 Kg
Internal resistance:	0.63 m Ohm
Short circuit current:	3250 A

Terminal: F-M8

Pole pairs: 1

The number of batteries connected in parallel is equal to the required (needed) battery capacity divided by the battery's nominal amp-hour. So,

Number of batteries connected in parallel = 482.48 Ah ÷ 550 Ah = 0.88. This means (with regard to batteries connected in parallel) that there is only **one** battery/battery string.

The nominal voltage of batteries connected in series is equal to the PV system's nominal voltage. Therefore, to find the number of batteries connected in series, divide the PV system's nominal voltage (24 VDC) by the battery's nominal voltage (2 VDC). So,

Number of batteries connected in series = 24V ÷ 2V = 12.

The total number of batteries is equal to the number of batteries connected in parallel (1) multiplied by the number of batteries connected in series (12). So,

Total number of batteries = 1 × 12 = 12

The PV system's nominal battery capacity is equal to the nominal capacity of the selected battery (550 Ah) multiplied by the number of batteries connected in parallel (1). So,

The PV system's nominal battery capacity = 550 Ah × 1 = 550 Ah

The usable battery (battery bank) capacity is equal to the PV system's nominal battery (battery bank) capacity multiplied by the battery maximum depth of discharge. So,

The usable battery (battery bank) capacity = 550 Ah × 0.80 = 440 Ah

Determining Design Electrical Current:

The winter tilt for the PV array is selected. Using sun-hours shown in the insolation data in **Table 8**, we see that the lowest sun-hours are during the months of July and August. We calculate the design electrical current for each month, at optimum fixed tilt angle year-round of latitude, of latitude minus 15^0 (favouring summer) and of latitude plus 15^0 (favouring winter). Thereafter, we select the value of the electrical design current and sun-hours for the period of high electrical current demand

and low insolation. Note that the selection of the design current will be such that the highest load demand does not unnecessarily make the PV system too big and costly.

The design current (at STC) is equal to amp-hours / day (Ah produced each day) divided by the sun-hours / day. The electrical design current obtained is at STC.

The electrical design current for each month for a PV array with a fixed tilt angle of latitude minus 15^0 year-round (fixed tilt angle for summer year-round) is calculated as follows:

- Design current for January = 128.66 amp-hours / day ÷ 5.92 sun-hours / day (for January) = 21.73A
- Design current for February = 128.66 amp-hours / day ÷ 6.07 sun-hours / day (for February) = 21.20A
- Design current for March = 128.66 amp-hours / day ÷ 5.43 sun-hours / day (for March) = 23.69A
- Design current for April = 128.66 amp-hours / day ÷ 4.89 sun-hours / day (for April) = 26.31A
- Design current for May = 128.66 amp-hours / day ÷ 4.60 sun-hours / day (for May) = 27.97A
- Design current for June = 128.66 amp-hours / day ÷ 4.18 sun-hours / day (for June) = 30.78A
- Design current for July = 128.66 amp-hours / day ÷ 3.36 sun-hours / day (for July) = 38.29A
- Design current for August = 128.66 amp-hours / day ÷ 3.70 sun-hours / day (for August) = 34.77A
- Design current for September = 128.66 amp-hours / day ÷ 4.57 sun-hours / day (for September) = 28.15A
- Design current for October = 128.66 amp-hours / day ÷ 5.06 sun-hours / day (for October) = 25.43A
- Design current for November = 128.66 amp-hours / day ÷ 5.60 sun-hours / day (for November) = 22.98A
- Design current for December = 128.66 amp-hours / day ÷ 6.16 sun-hours / day (for December) = 20.89A
- Yearly average design current = 128.66 amp-hour / day ÷ 4.96 sun-hours / day = 25.94A

Similarly, we calculate the electrical design current for each month for a PV array with fixed tilt angle of latitude year-round. So,

- Design current for January = 128.66 amp-hours / day ÷ 5.56 sun-hours / day (for January) = 23.14A
- Design current for February = 128.66 amp-hours / day ÷ 5.87 sun-hours / day (for February) = 21.92A

- Design current for March = 128.66 amp-hours / day ÷ 5.49 sun-hours / day (for March) = 23.44A
- Design current for April = 128.66 amp-hours / day ÷ 5.19 sun-hours / day (for April) = 24.79A
- Design current for May = 128.66 amp-hours / day ÷ 5.11 sun-hours / day (for May) = 25.18A
- Design current for June = 128.66 amp-hours / day ÷ 4.75 sun-hours / day (for June) = 27.09A
- Design current for July = 128.66 amp-hours / day ÷ 3.71 sun-hours / day (for July) = 34.68A
- Design current for August = 128.66 amp-hours / day ÷ 3.95 sun-hours / day (for August) = 32.57A
- Design current for September = 128.66 amp-hours / day ÷ 4.68 sun-hours / day (for September) = 27.49A
- Design current for October = 128.66 amp-hours / day ÷ 4.97 sun-hours / day (for October) = 25.89A
- Design current for November = 128.66 amp-hours / day ÷ 5.31 sun-hours / day (for November) = 24.23A
- Design current for December = 128.66 amp-hours / day ÷ 5.72 sun-hours / day (for December) = 22.49A
- Yearly average design current = 128.66 amp-hours / day ÷ 5.03 sun-hours / day = 25.58A

Similarly, we calculate the electrical design current for each month for a PV array with fixed tilt angle of latitude plus 15^0 (fixed tilt angle for winter year-round). So,

- Design current for January = 128.66 amp-hours / day ÷ 4.94 sun-hours / day (for January) = 26.04A
- Design current for February = 128.66 amp-hours / day ÷ 5.40 sun-hours / day (for February) = 23.83A
- Design current for March = 128.66 amp-hours / day ÷ 5.30 sun-hours / day (for March) = 24.28A
- Design current for April = 128.66 amp-hours / day ÷ 5.27 sun-hours / day (for April) = 24.41A
- Design current for May = 128.66 amp-hours / day ÷ 5.42 sun-hours / day (for May) = 23.74A
- Design current for June = 128.66 amp-hours / day ÷ 5.14 sun-hours / day (for June) = 25.03A
- Design current for July = 128.66 amp-hours / day ÷ 3.93 sun-hours / day (for July) = 32.74A
- Design current for August = 128.66 amp-hours / day ÷ 4.04 sun-hours / day (for August) = 31.85A

- Design current for September = 128.66 amp-hours / day ÷ 4.60 sun-hours / day (for September) = 27.97A
- Design current for October = 128.66 amp-hours / day ÷ 4.66 sun-hours / day (for October) = 27.61A
- Design current for November = 128.66 amp-hours / day ÷ 4.77 sun-hours / day (for November) = 26.97A
- Design current for December = 128.66 amp-hours / day ÷ 5.02 sun-hours / day (for December) = 25.63A
- Yearly average design current = 128.66 amp-hours / day ÷ 4.87 sun-hours / day = 26.42A

The values of electrical design current and sun-hours for the period of highest electrical current demand and low insolation are obtainable during the months of July and August. In order to enable the PV system to function efficiently it is necessary to select the design month as the month during which the design current is highest and the sun-hours are lowest. Since the PV system availability is expected to be below 99% (PV system powering household), the load is not critical. Therefore, energy management will be carried out during the months of July and August.

The selected watt-hours per day for refrigerator and freezer, with a duty cycle of nine hours per day and five hours per day, respectively, are for summer energy use. During winter months, the refrigerator and freezer have a duty cycle of seven hours per day and four hours per day, respectively. Therefore, this will reduce the peak energy demand for July and August. Additionally, the fan will not be utilised in winter. The operation of the PV system will also be monitored and, if further reduction in energy demand is required, the duty cycle for the television will be reduced from four hours to three hours per day.

The highest design current of 38.29A for the PV array at the summer tilt angle (fixed at latitude minus 15^0 tilt), for the month of July, is 12.35A more than the yearly average design current (38.29 – 25.94 = 12.35A). The highest design current of 34.68A for the PV array at latitude tilt angle (fixed at latitude tilt) is 9.1A more than the yearly average design current (34.68A – 25.58A = 9.1A). The highest design current of 32.74A, for the PV array at winter tilt angle (fixed at latitude plus 15^0 tilt), is 6.32A more than the yearly average design current (32.74A – 26.42A = 6.32A).

Therefore, the highest design current of 38.29A, for the fixed summer tilt angle (fixed at latitude minus $15\ ^0$ tilt) of the PV array, makes the PV system unnecessarily too big and costly. The same can be said of the PV array's fixed tilt at latitude. In addition, both the fixed summer tilt (latitude minus 15^0 tilt) and latitude tilt have shallower tilt angles compared to the fixed winter tilt (latitude plus 15^0). Since the fixed tilt at latitude plus 15^0 (winter tilt) has the largest tilt angle, the cleaning of the PV array is much easier. Of the winter months we select the month with the highest demand and lowest insolation.

Therefore, the design current of 32.74A, with corresponding sun-hours of 3.93, is adopted. The design current does not take into account the PV module de-rate factor. The PV array's tilt angle is equal to latitude plus 15^0 (which is the winter tilt year-round). So,

PV array tilt angle = $8.82^0 + 15^0 = 23.82^0$. This means that the PV array tilt angle will be 25^0 (approximately).

Determine PV system array size:

The technical data sheet for the selected PV module will include the following (voltages and currents are at standard test conditions):

• PV module tolerance ± 3%	
• Voltage at maximum power point (V_{MMP})	30V
• Current at maximum power point (I_{MMP})	7.92A
• Open-circuit voltage (V_{OC})	37V
• Short circuit current (I_{SC})	8.31A
• Output voltage (for charging the battery)	18V
• Output voltage at highest site temperature	15.07V
• Weight	21kg
• Dimensions	1640mm x 990mm x 50mm
• Operating temperature	-40^0 to $+85^0$

The battery charging voltage is equal to the battery's nominal voltage multiplied by the number of batteries in series and by a factor of 1.2. So,

Battery charging voltage = 2V × 12 × 120% = 28.8V

The number of PV modules in series is equal to the battery charging voltage divided by the adjusted PV module output voltage at the highest PV module temperature at the site. So,

Number of PV modules in series = 28.8V ÷ 15.07V = 1.91 (this should always be rounded up). This means that the number of PV modules in series will be equal to 2.

Usually, PV modules do not operate at ideal conditions. Therefore, it is necessary to apply the PV module de-rate factor, due to PV module tolerance and/or dirt (e.g. dust) accumulation on the PV module's surface, to obtain the required design current. Since the modules will be regularly cleaned, the factor for dirt accumulation on the PV module's surface is taken as 0.98. Power loss due to PV module tolerance (de-rate factor) is 0.97 (PV module technical data sheet). Therefore, the total PV module de-rate factor is 0.97 × 0.98 = 0.95.

Adjusted (de-rated) design current (current produced by the PV module when it is operating at 1000W / m^2, 25^0C, and AM1.5) is equal to the design current divided by the PV module de-rate factor. This is the current obtainable at ideal conditions. So,

De-rated design current = 32.74A ÷ 0.95 = 34.46A

The number of PV modules in parallel is equal to the de-rated design current divided by the PV module rated current at maximum power point (I_{MPP}). So,

The number of PV modules in parallel = 34.46A ÷ 7.92A = 4.35 PV modules. This means that the number of PV modules in parallel will be equal to 4.

The total number of PV modules is equal to the number of PV modules in series multiplied by the number of PV modules in parallel. So,

The total number of PV modules = (2 × 4) PV modules = 8 PV modules.

The PV array current, at maximum power point, is equal to the number of PV modules in parallel multiplied by the rated PV module current (I_{MPP}) at standard test conditions. So,

PV array current (I_{MP}) = 4 × 7.92A = 31.68A

The PV array short-circuit current (I_{SC}) is equal to the number of PV modules in parallel multiplied by the PV module rated short-circuit current at standard test conditions ($I_{SC\ STD}$). So,

PV array short-circuit current ($I_{SC\ STD}$) = 4 × 8.31A = 33.24A

The PV array open-circuit voltage is equal to the number of PV modules per PV string multiplied by the PV module open-circuit voltage at standard test conditions ($V_{OC\ STD}$). So,

PV array open-circuit voltage ($V_{OC\ STD}$) = 2 × 37V = 74V

Charge controller input current should not be less than the PV array short-circuit current (STD) multiplied by the factor of 1.25 as a minimum (***6.5.3.1 Selecting battery charge controller***). So,

Minimum charge controller input current = 33.24A × 1.25 = 41.55A

Charge controller input voltage should not be less than the PV string open-circuit voltage (STD) multiplied by a correction of 1.04 as a minimum (***BS IEC 62548: 2016 - Photovoltaic (PV) arrays, section 7.2, Table 4***). So,

A decision has been made to divide the PV array into two parts in order to install cheaper controllers rated 30A or less. Each part of the PV array will have a separate charge controller. The controllers will be connected in parallel to each other and connected together at a point before the battery bank. The minimum current for each charge controller should be equal to half the number of PV modules connected in parallel in the PV system array multiplied by the PV module short-circuit current.

Minimum charge controller current = 41.55A ÷ 2 = 20.78A

Minimum charge controller input voltage = 74V × 1.04 = 76.96V.

Select battery charge controller:

The rated current of the selected battery charge controller is 30A.

Specification of the selected battery charge controller includes:

Model:

Nominal system voltage (24V);

Rated current (30A);

Charging array current indication;

Charging voltage indication;

Low voltage warning indication;

Equalize charging voltage;

Boost charging voltage;

Float charging voltage;

High-voltage-disconnect (HVD) and high-voltage-reconnect;

Low-voltage-disconnect (LVD) and low-voltage-reconnect;

Temperature compensation;

Terminals ($16mm^2$);

Working temperature (-35^0 to $+55^0$).

Determining PV array to load ratio (see ***6.7.1.3 Preliminary design and electrical energy estimate***):

W-h / day = 2783.94 W-h

The de-rated current is equal to 34.46 Ah, and the PV system voltage is equal to 24VDC

So,

PV array power = 34.46A × 24V = 827.04W-h

PV array to load ratio = 827.04W-h ÷ 2783.94W-h = 0.297

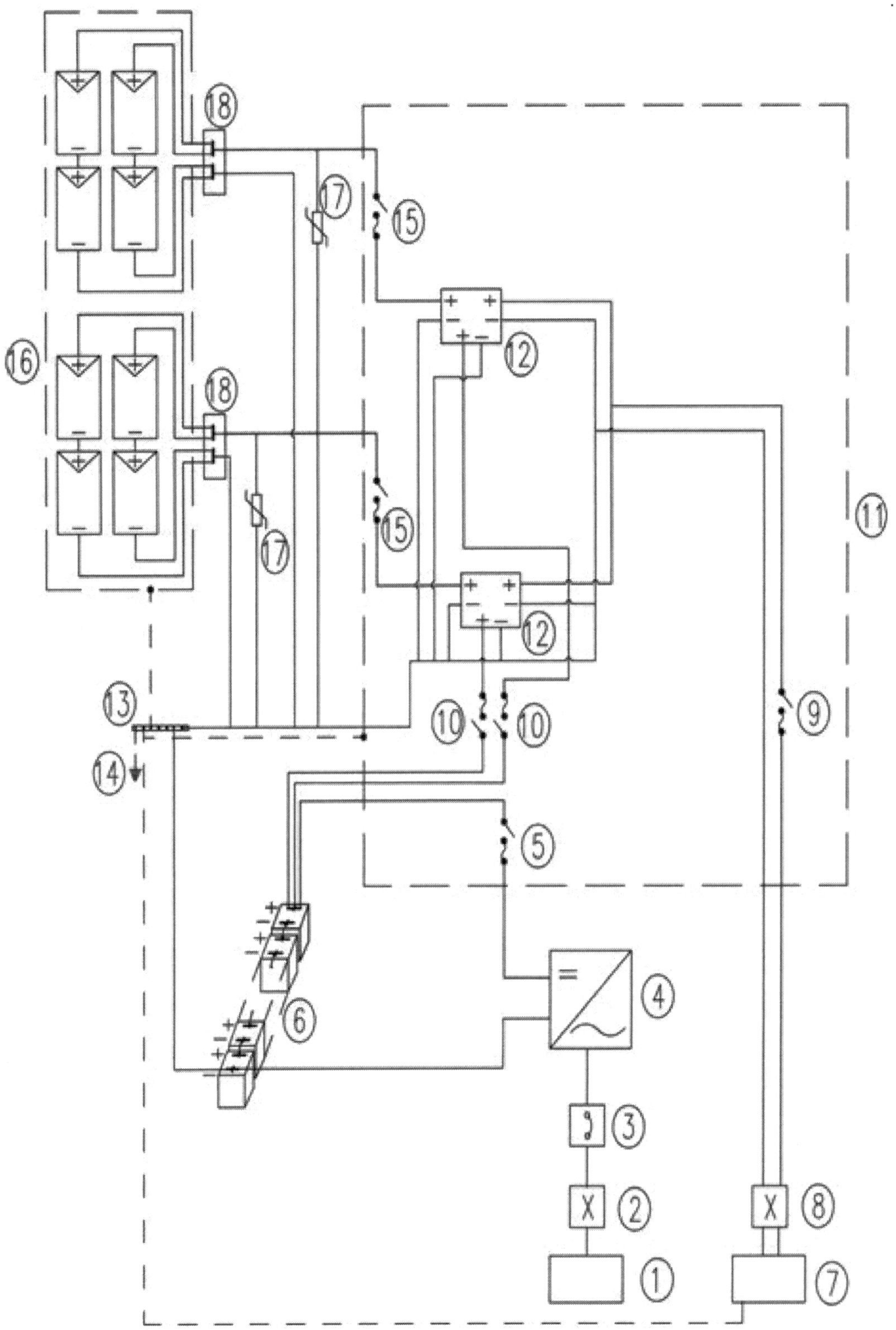

Figure 25: Standalone PV system layout

Table 6: This table shows notes to Fig. 25

REF	DESCRIPTION
1	DISTRIBUTION BOARD (AC LOADS) IN KITCHEN
2	AC DOUBLE POLE ISOLATOR
3	CIRCUIT BREAKER
4	INVERTER (WALL-MOUNTED), IN GARAGE
5	SWITCH FUSE
6	BATTERY BANK, ON BATTERY RACK, IN BATTERY ROOM
7	DISTRIBUTION BOARD (DC LOADS), WALL-MOUNTED, IN KITCHEN
8	DC DOUBLE POLE ISOLATOR
9	SWITCH FUSES
10	SWITCH FUSES
11	CONTROL CENTRE BOX, WALL-MOUNTED, IN GARAGE
12	CHARGE CONTROLLERS
13	NEGATIVE BUS
14	EARTH ROD
15	SWITCH FUSE
16	PV ARRAY
17	SURGE PROTECTOR (IN PV CONTROL CENTRE BOX)

7. Reference Material

For further information see the references given elsewhere in this book and see the list below:

1. *Solar Power magazine*
2. *Solar Electricity*, second edition, in UNESCO's Engineering Series.
3. *Photovoltaics for Professionals: Solar Electric Systems, Marketing, Design and Installation.*
4. *Standalone Photovoltaic Systems: A Handbook of Recommended Design Practices*, issued by Sandia National Laboratories, USA.
5. *Progress in Photovoltaics*, Solar Cell Efficiency Tables (Version 48) – Accessed 2017.
6. *B. Y. H. Lui and R. C. Jordan (1960), "The interrelationship and characteristic distribution of direct, diffuse and total solar radiation," solar energy vol. 4 no. 3, pp 1-19, 1960.*
7. *Charles R. Landau (2015), "Optimum Tilt of Solar Panels," Accessed 2017. Htt://www.solarpaneltilt.com*
8. *J. A. Duffie and W. A. Beckman (1991), Solar Engineering of Thermal Process, John Wiley& Sons, New York, NY, USA.*
9. *B. Y. H. Lui and R. C. Jordan (1962), "Daily insolation on surfaces tilted towards the equator," Trans ASHRAE 67:526 – 541.*
10. *Heywood (1971), "Operating experience with solar water heating," IHVE J, vol. 39:63 – 69.*
11. *G. Lewis (1987), Optimum tilt of solar collector," Solar and Wind Technology, VOL. 4, no. 3, pp. 407-42.*
12. *H. P. Garg (1982), "Treatise on Solar Energy," Vol. 1 of fundamentals of solar Energy, Wiley, New York, NY, USA.*
13. *Jose Cano (2011), "Photovoltaic Modules: Effect of Tilt Angle on Soiling," Accessed 2017. https://repository.asu.edu/attachments/57216/contents/Cano _asu_0010N 11001.pdf*
14. *Riordan and Hulstron (1990), "What is an air mass 1.5 spectrum? Solar cell performance calculation," Pages: 1085 – 1088, vol.2.*
15. *J. I. Jimenez and Y. Castro (1982), Solar radiation on sloping surfaces with different orientations in Granada, Spain," Solar energy, 28, 257 - 262*
16. *T. Muneer (2004), Solar Radiation and Daylight Models, Elsevier, Berlin, German, 2nd edition.*
17. *P. S. Koronakis (1986), "On the choice of tilt angles for south facing solar collectors in the Athens Basin Area," Solar Energy 33:217 – 225.*
18. *V. Badescu (2002), "3D isotropic approximation for solar diffuse irradiance on tilted surfaces," Renewable Energy 26:221 – 223.*

19. *J. E. Hay and J. A. Davies (1980), "Calculation of monthly mean solar radiation for horizontal and inclined surfaces," Solar energy, 23, 301 – 307.*

20. *D. T. Reindel at al. (1990), "Evaluation of hourly tilted surface radiation models," Solar Energy 45, 97- 17.*

21. *C. V. Nayar et al. Novel wind / diesel / battery hybrid energy system, Sol Energy, 51 (1): 65-78.*

8. About The Author

Phineas S. Malungwa was born on 29th of March 1943, in Nyamandlovu District, near Bulawayo, Zimbabwe.

Between 1980 and 1985, Phineas did electrical engineering at The Institute for Mechanical & Electrical Engineering, Varna, Bulgaria. In 1985 he was awarded a "Diploma in Electrical Power Supply Systems & Electrical Equipment", equivalent to Masters of Science degree (Electrical Eng.). Solar electricity was one of the subjects.

As a building services electrical engineer, his designs included the following systems: schematics, power and data points, lighting and emergency lighting, fire alarm, lightning protection, photovoltaic.

In 2006 – 2007, Phineas worked for Capita Symonds as senior electrical engineer, he was responsible for design and supervision of construction projects, and also carried out regular inspections of completed projects.

In Feb 2001, Phineas started a consulting company, under the name "P.S. Malunjwa & Co. Consulting Building Services Engineers". In 2001 to 2004, he worked for P.S. Malunjwa & Co. Consulting Building Services Engineers as senior electrical engineer responsible for electrical designs and supervision of construction projects. The largest projects were two super markets.

In 1999 – 2000, he worked for Network Building Services Engineers, as senior electrical engineer and was responsible for electrical designs and supervision of construction projects. The biggest project was the teacher training college with power supply of 2 mega-watts.

In 1996 – 1999, Phineas worked for Department of Public Works, Zimbabwe and was employed as Provincial Officer Operations (equivalent to chief electrical engineer). His work involved the administration of the Department of Public Works in the province, assisting junior offers, supervision of all construction works and maintenance of government buildings in the province, and assisting local authority and nongovernmental organisations on technical issues.

In 1995 – 1996, he worked for Department of Public Works, Zimbabwe and was employed as Deputy Provincial Officer Operations (equivalent to principal electrical engineer). His work involved the administration of the Department of Public Works in the province, assisting junior offers, supervision of all construction works and maintenance of government buildings in the province, and assisting local authority and nongovernmental organisations on technical issues.

In 1990 – 1995, Phineas worked for Department of Public Works, Zimbabwe as a design engineer, responsible for electrical designs and supervision of construction projects, which included mainly, residential houses, primary and secondary schools, youth training centres, technical colleges, office blocks. The design and supervision were for both conventional electricity and solar electricity. The solar electricity was mainly for clinics and homes in remote areas. In 1994 he was elevated to senior electrical engineer.

In 1990 – 1995, he worked for ministry of education as O' level school teacher for mathematics and science. Also, between 1989 – 1990, Phineas was employed (part-time) as lecturer on low voltage networks, and electromagnetic, for second and first degree students (for Bachelor of Technology Programme, University of Zimbabwe).

From 12 to 16th, Phinea participated in a UNESCO sponsored sub-regional workshop, entitled "Renewable Energy Sources", held at SIRDC (Scientific Industrial Research & Development Centre), Harare, Zimbabwe.

In addition Phineas has an interest in poetry writing, he wrote his first English poetry book called "The Cry for Freedom Justice and Peace". He had written several poems before, in vernacular (Ndebele language, which is his mother tongue). He says that his best poems in vernacular are in a poetry book called "Umdumo Wezinkondlo", which literary means "the thunder of poems". This poetry book is a poetry text book for secondary schools and colleges in Zimbabwe.

Phineas is also interested in watching sports, especially athletics and football. He was a very strong competitor in athletics (220 to 800 metres) inter-schools competitions in the country, in his youthful years.

ISBN 978-1-912505-31-9

Made in the USA
Monee, IL
07 May 2026

49714800R00074